Security Informatics and Law Enforcement

Series Editor

Babak Akhgar, CENTRIC (Centre of Excellence in Terrorism, Resilience, Intelligence and Organised Crime Research), Sheffield Hallam University, Sheffield, UK

The primary objective of this book series is to explore contemporary issues related to law enforcement agencies, security services and industries dealing with security related challenges (e.g., government organizations, financial sector insurance companies and internet service providers) from an engineering and computer science perspective. Each book in the series provides a handbook style practical guide to one of the following security challenges:

Cyber Crime - Focuses on new and evolving forms of crimes. Books describe the current status of cybercrime and cyber terrorism developments, security requirements and practices.

Big Data Analytics, Situational Awareness and OSINT - Provides unique insight for computer scientists as well as practitioners in security and policing domains on big data possibilities and challenges for the security domain, current and best practices as well as recommendations.

Serious Games – Provides an introduction into the use of serious games for training in the security domain, including advise for designers/programmers, trainers and strategic decision makers.

Social Media in Crisis Management – explores how social media enables citizens to empower themselves during a crisis, from terrorism, public disorder, and natural disasters

Law enforcement, Counterterrorism, and Anti-Trafficking – Presents tools from those designing the computing and engineering techniques, architecture or policies related to applications confronting radicalisation, terrorism, and trafficking.

The books pertain to engineers working in law enforcement and researchers who are researching on capabilities of LEAs, though the series is truly multidisciplinary – each book will have hard core computer science, application of ICT in security and security/policing domain chapters. The books strike a balance between theory and practice.

Fraser Sampson

How AI-Driven Surveillance is Changing our World

Do Robots Have the Right to Remain Silent?

Springer

Fraser Sampson
CENTRIC (Centre of Excellence
in Terrorism, Resilience, Intelligence
and Organised Crime Research)
Sheffield Hallam University
Sheffield, UK

ISSN 2523-8507 ISSN 2523-8515 (electronic)
Security Informatics and Law Enforcement
ISBN 978-3-032-24004-0 ISBN 978-3-032-24005-7 (eBook)
https://doi.org/10.1007/978-3-032-24005-7

This Springer imprint is published by the registered company Springer Nature Switzerland AG
The registered company address is: Gewerbestrasse 11, 6330 Cham, Switzerland

Preface

The use of AI-driven surveillance technology—in particular biometrics like facial recognition—by the state is one of the most contentious issues in law enforcement across the world. While some argue that if you have done nothing wrong you have nothing to fear, others point to irreversible erosions of privacy and a dystopian Orwellian future. Whether you are a policymaker or police leader, influencer or academic or simply one of a growing number of concerned citizens, the technology is being used extensively, sometimes in ways that have escaped our attention. AI-enabled surveillance has already changed the relationship between the citizen and the state and the traditional predictions of what the future will look like.

Drawing upon his direct professional experience from policing, legal practice and UK regulator, the author takes complex, profound and highly topical themes and reduces them to simple but important questions as the title implies. In a collection of highly readable and relatable essays, the author accompanies the reader through everyday examples of the growing use of AI-driven surveillance, challenging received wisdom and pointing out the challenges for law enforcement and the citizen along the way.

At its heart, this is a guidebook charting the technological, legal and societal implications of previously unimagined surveillance technology, not only in the hands of state agencies, but principally in our own.

Sheffield, UK Fraser Sampson

Contents

1 Are AI-Driven Devices Simply a Question of Evolution?

Since the dawn of technology, we were on speaking terms with our tools. So said technology author George Dyson in his seminal work *Analogia* (Dyson, 2020) where he maps our circular evolution from flint knappers to micro chippers. In making the evolutionary leap from adze to Azure, it's possible that notions of kinship might be one of the reasons behind our early instincts that all Artificial Intelligence (AI) must be 'explainable' (Yang et al., 2023). There are acres of text about the importance of AI's explainability, ranging from heavyweight academic treatises to legal guidance and public policy—far too many to reference here. But the assumption that all other technologies in common use are immediately and fully understood by everyone using them is simply not borne out by historical evidence. We haven't been acquainted with many of our tools for a while and you don't need to go back to the dawn of technology to see that—somewhere around mid-morning should do.

When people unpacked their first Macintosh home computer, how many understood the workings of its revolutionary graphical user interface? My parents certainly didn't. As computer pointers evolved from CAT to mouse, who among the uninitiated mass of domestic users acquainted themselves with the technological genius of the emerging tools under their fingers? With the arrival of the touchpad I'm using to write this chapter there's no need for me to grasp either the device or its theory. Whether Dyson is right about our ancestral familiarity with our implements, there has been no demand for explicability in the workings of our other digital tools—we have been ready, willing and able to accept with gratitude the ingenuity that's letting us work remotely, cheaply and with increasing frequency. Why then would AI-derived tools be any different?

F. Sampson, *How AI-Driven Surveillance is Changing our World*, Security Informatics and Law Enforcement, https://doi.org/10.1007/978-3-032-24005-7_1

There are several potential reasons but let's start with a similarity. One thing about AI that is no different from previous eras is the atavistic fear it has stirred. Some debates about generative AI's very existence for example (see e.g. Eisikovits, 2023) could just as easily be about witchcraft.

Historical headlines portending how we would use the new 'telephone' to communicate with the dead (Stephens, 1998) or alarm about the psychological impact of radio on children (Orben, 2020) show there's nothing new about neophobia. Even when we're tapping in all caps with our opposable thumbs, our tendency to greet technological innovation with alarm is corroborated by history. Enduring ancient superstitions aside, there are however several empirically sound reasons why the arrival of AI might need specific explanation in a way that didn't apply to discoveries of the past. The first of these is evolution itself—not ours but that of our tools.

Whether or not humanity has enjoyed a deep understanding with our industrial machinery so far, this time around the tools will be wholly different. They are already able to 'evolve' without (or even despite) us. This may not (yet) be recognised or labelled as evolution in the traditional biological sense, but that simply illustrates the novelty of the territory we are now in. Our lexicography will need to accommodate AI, and we're going to need a bigger dictionary. Whatever we choose to call it, anything that possesses the aptitude for self-perpetuation and imitation deserves explanation.

Another feature of AI-enabled tools is our relevance. Tools from every era since Dyson's 'dawn of time' have needed us—their creators—to direct, adjust and improve them. Those of the next era will not. They are already able to teach themselves and each other beyond the narrow parameters of their initial programming. That's an astounding difference, a difference so elemental that AI-enabled 'tools' deserve names that better reflect their speciose nature (as opposed to just their genus like Large Language Model—LLM) like *hominins*. Autodidactic, idiolectal entities are not a 2.0 version of any implements that went before. They're not a discovery of a new element to be fixed into the periodic table. They're a wholly new category of existence and grappling with the enormity of its implications will call for more than digital dexterity on our part. Why should Artificial Intelligence have to explain itself? Because it can.

A further reason why AI-enabled capabilities will need to be explainable is automated decision-making. Thus far, humanity's use of tools to 'make decisions' has been heavily regulated and restricted by legal frameworks such as the General Data Protection Regulation (GDPR) and the EU AI Act of the European Union.[1]

[1] https://gdpr-info.eu/ art. 22.

The fact that early applications of AI systems were often described as utilising methodology and processes that even the designers were unable to explain has engaged our innate suspicions (above) and reinforced the case for explanation. AI algorithms built on machine learning systems are 'in many cases, functionally black boxes to humans' (Bathaee, 2018). With consequential tools that are having a real-life impact on our lives, we need to know how those decisions are made, not simply to understand them, but also if we're to able to challenge them. The computerisation of administrative and even some executive functions has been around for several decades. Where those computational interventions take place in a police setting like surveillance or investigative recordkeeping, the implications are already significant[2] When the tools used by the state for law enforcement, distribution of benefits, control of borders and health, etc., clear, intelligible explanation will be essential to the acceptance of both tools and their outputs. Surveillance tools of the past—like CCTV[3]—haven't needed to 'show their working out in the margin' (other than to auditors): tools capable of being *calculating* rather than simply calculators will be accountable.[4] Inexplicable objects producing mechanical outcomes are just another form of bewitchment which the citizen may accept in fiction (think of the Sorting Hat of Harry Potter stories) but not in real life.

One response to the potential tyranny of the black box has been to insist on human agency. This might be another link back to our first fear and mistrust of machines but, whatever its provenance, the requirement for a so-called human-in-the-loop is enshrined in legislative frameworks in many countries[5] and is standard for AI policies. Insistence on human oversight is another reason why AI-enabled tools are different from anything that has gone before and why they must be explained. If the human on whom we are pinning so much is to be able to step in and override its decisions, then they had better be on more than just speaking terms with the AI enabled tools.

These are all credible reasons why AI-enabled capabilities must be explainable and explained. And the need for explanation/intervention is probably at its greatest where the tools are used by the state to make decisions about us, our lives, liberty and livelihoods. A public space surveillance example illustrates this point.

[2] And are one the reasons behind the Worst Surveillance Advice Ever—see Chapter 9.

[3] Chapter 5.

[4] Chapter 10.

[5] See, e.g., GDPR.

Whether or not drivers have understood them, speed cameras have been monitoring our driving for decades. Roadside and gantry-rigged cameras in the UK collect in excess of 80 million number plate 'reads' every day, tracking a vehicle's credentials and feeding the data into a national network covering roads, car parks and other public places. AI-enabled road safety tools will look and operate very differently from these standard cameras. They will increasingly be 'on speaking terms' with our in-car telematics which means they will have far more information than just the visible and legible number plate. They will be able to testify as to what speeds we've been doing for the last year not just the last mile. They will have a good idea who the driver is, be able to predict their next journey, likely passengers and interaction with any other vehicles. They will eventually be permitted to be on speaking terms with juries, courts and public inquiries, reconstructing events that took place before they were invented, together with all the sounds and smells and sensation rather than just producing pictures and statistics. Much will depend on the type of AI involved but even the *genus* currently known as Narrow AI has given the police phenomenal new surveillance reach as the remainder of this book will show. Anything that can reduce injury, death and delay on a country's road networks will attract considerable political and societal support, but it will have to meet new standards and tests of accountability. That means explanation.

The use of AI-enabled biometrics like facial recognition technology[6] remains controversial in law enforcement but is becoming more prevalent in the UK. Some of that controversy has come from clumsy experimentation; some of the expansion has come from better explanation. By taking the time and investing effort in explaining how the technology works, what it is (and is not) doing and how it meets technical, legal and societal expectations, the police have begun to unlock some of the extraordinary potency offered by remote biometrics. The ability to use AI-enabled tools to match historical data from pictures of faces in stills or video captures for purposes of elimination and implication presents the state with a new epoch in surveillance. But it is the *prospective* decision-making power of AI that holds the real payoff for cost-effectiveness and scalable impact in crime prevention, investigation and management. In a law enforcement context, knowing where, when and how incidents are likely to arise will transform capability and deployment of resources. AI-enabled surveillance capabilities will accelerate actuarial policing like law enforcement's own Large Hadron Collider (minus the 10,000 scientists and ten-year project cycle). If we can predict accidents, prevent incidents and pre-empt criminality, saving time, money and lives

[6] Chapter 6.

using AI, wouldn't it be morally delinquent not to? But harnessing that power will need hypervigilant governance accompanied by unequivocal explanation. Recent research at CENTRIC[7] has revealed some very interesting attitudes among citizens, both in the UK and across other nations. In one large survey, a majority of citizens across the UK said they support the use of AI in policing, provided that it was accompanied by strong accountability, oversight and safeguards.

The survey consulted 10,114 citizens across all four UK nations, and respondents ranged in age from 14 to 85+ years. The overall results found positive perceptions of the police use of AI tools, with 61.3% of citizens supporting the use of AI by the police and 64.0% believing that it enhanced police capabilities (compared to 12.4% who believe AI in fact undermines them).

Respondents' core concerns were a **lack of accountability, the** potential for 'oversurveillance', misuse of their personal data and the prospect of the police 'blaming' AI for their errors. While 61.3% expressed general support for the police using AI-enabled technology, nearly 40% said they would be unwilling to contribute personal data for the training of police AI systems and only 9.3% would be very willing to do so.[8].

Citizens' views were also heavily dependent on the specific context in which the police use the AI tools, something that will be particularly relevant in the catch-all surveillance capabilities that AI-enabled technology is bringing. From a choice of 32 different deployment scenarios, respondents' trust and acceptance of the AI tended to be highest for situations carrying low risk for the public and a**cceptance was lowest for use that some might see as relatively uncontentious, such as the automation of core functions** like emergency call handling.

Perhaps most importantly of all, was the public's expectation of consistent oversight and accountability. Three quarters of respondents wanted mandatory accountability processes in place *before* the police use AI-enabled tools and only one fifth believe that sufficient accountability currently exists.

A separate, international survey on the accountable use of AI by the police reported that nearly three in four participants across 30 countries agreed that the police should use AI to predict crimes before they happen (74.3%). That's not nearly as straightforward as it sounds.[9] It is a very short—and perhaps irresistible—evolutionary step for those same tools to try to identify, not only the offence, but also the future offender. In the first survey, the highest acceptance

[7] Centre of Excellence in Terrorism, Resilience, Intelligence and Organised Crime Research at Sheffield Hallam University.

[8] This is part of the 'policing privacy paradox'—see Chapter 18.

[9] As explained in Chapter 10.

levels among UK citizens were given where the technology was to be used for identifying 'potential' perpetrators of child sexual exploitation and terrorism. That way lies an unexplored galaxy of dark matter.

The final reason AI-enabled tools need explanation is because their existence upsets our ancestral dominance. Dyson concluded that we were heading for 'a world where humans coexist with technologies they no longer control'. Perhaps AI's greatest difference from what's gone before lies here. With the arrival of AI-enabled capability, we are moving into the first power sharing arrangement with our tools, replacing our absolute historical authority over them. This brings the prospect that we may one day have to explain ourselves to them.

If we have always enjoyed an intimate relationship with our tools in the past, we are probably the last generation where that holds good. As our tools have proliferated, we have regressed and—as becomes alarmingly apparent when any of them malfunction—are now barely on nodding terms with devices that sustain our everyday lives.

Being 'on speaking terms' describes a reciprocal preparedness to engage. On one view, ours is in fact the *first* generation to be on literal speaking terms with its tools—and the first since the dawn of technology where our tools may not be on speaking terms with us. That requires explanation.

References

Bathaee, Y. (2018). The artificial intelligence black box and the failure of intent and causation. *Harvard Journal of Law and Technology, 31*(2), 890–938.

Dyson, G. (2020). *Analogia: The entangled destinies of nature, human beings and machines.* Allen Lane.

Eisikovits, N. (2023, July 12). AI is an existential threat—Just not the way you think. *The Conversation.*

Orben, A. (2020). The Sisyphean cycle of technology panics. *Perspectives on Psychological Science, 15*(5), 1143–1157.

Stephens, M. (1998). *The rise of the image, the fall of the world.* Oxford University Press.

Yang, W., Wei, Y., Wei, H., et al. (2023). Survey on explainable AI: From approaches, limitations and applications aspects. *Human-Centric Intelligent Systems, 3*, 161–188.

2 Why Would China Want to Hack my Fridge?

According to the traditional song, he knows if we've been bad or good—perhaps that explains our seasonal obsession with surveillance stories. In a later chapter we will look at another diarised surveillance favourite—the end-of-year listing of the world's most spied-on cities. For now, let's consider the impact of AI-enabled technology through the lens of seasonal giving.

Stories about hostile states using presents to spy on us have become something of a festive tradition but we should remember that tales dogging surveillance technology are not just for Christmas. There are traditionally more bad tidings at that time of year—who can forget those innocent-eyed little Furbies, charged (though never convicted) with spying for China back in the Christmas of '98?.[1] But is all this 'spy cams under the tree' just gift-wrapped surveillance clickbait or is there something here that we should take seriously?

With the rise of the seasonal smart device, every year the Internet of Things (IoT) produces enough headlines to fill the Twelve Days of Christmas (ten spies a-watching, nine watches spying, eight doorbells dobbing, etc.), but the broader topic of hacking domestic appliances like dishwashers and freezers is perennial.[2]

All the carolling about air fryers cooking our digital goose, and knitting machines stitching us up, is manna from heaven for stand-up comedy, but dismissing the stories out of hand misses a key point. The stories have, as all good stories must, a kernel of truth and here provide a good context to look at a few truths about global surveillance in the age of AI. It is easy to laugh off warnings

[1] https://melmagazine.com/en-us/story/original-furby-1998.

[2] https://www.thetimes.com/uk/defence/article/embedded-chinese-tech-could-freeze-cars-and-traffic-lights-fw8hd20v0.

F. Sampson, *How AI-Driven Surveillance is Changing our World*, Security Informatics and Law Enforcement, https://doi.org/10.1007/978-3-032-24005-7_2

about state-sponsored hacking of our domestic appliances. You may hear people chuckling about the absurdity of a state hoping to extract something valuable from the dreary content of their white goods: 'any government that wants to look inside my washing machine …' [Fill in your own punchline].

For those who lived through the Cold War, a foreign power infiltrating parts of our establishment with spies and agents is something that we have grown up with. For others, the storied experience of espionage via Hollywood and Netflix has generally been built around themes of infiltration, expatriation and assassination. Developments in technology mean our adversaries' cyber capabilities now extend across any geographical boundaries and they can break into bank accounts to syphon funds, or hack computers to drain data. In those contexts, the payoff makes sense, but what would another country gain by accessing your TV or tumble drier? The answer is not really about surveillance at all, or at least not *just* about the surveillance and who's looking or listening. One of the principal risks underlying these stories comes less from what can be seen and more from what can be sown.

Police officers know that one of the most distressing things about burglary is the feeling of violation that it creates. The intrusive nature of 'breaking and entering' can have a significant psychological impact on the victim, scaring and scarring them, both at the time and afterwards the event, leaving behind a lingering feeling of insecurity. That is why the criminal laws in almost every jurisdiction on Earth treat breaking into, and stealing from, a home more seriously than other types of theft. Imagine then the impact if the intruder was not a shadowy masked individual but an invisible foreign government.

If a hostile state could reach into your home and rummage through your life at will, shutting down the router or dialling up the thermostat right under your nose—and your government's—that wouldn't just be unsettling *poltergeistery*. On an individual level, that would be an incursion into some of our most elemental places of safety and security. At a community level, distributed domestic disruption by another country would be a silent and sinister demonstration that we are within easy range. And at an international level, it would be cyber warfare. Officials have long warned us about the ability to 'wreak havoc' with communications services providers, manufacturers, utilities suppliers and government bodies.[3] But subtler, domestic interference is a potent threat and it is not just a traditional chestnut to be mulled over once a year. It is a constant risk and requires perpetual vigilance.

[3] https://www.ncsc.gov.uk/news/uk-allies-expose-china-tech-companies-enabling-cyber-campaign.

We have supercharged the extent of this threat by fetishising remote gadgetry and increasing the number of smart devices in our lives that we then come to rely on.[4] Our evolution from 'analogue humanity to digital humanity' (Laterza, 2021) has not only brought an eagerness to share detailed accounts, searchable images and minutely referenced data points relating to every aspect of our lives (Sampson, 2026), but also a determination to connect everything and control it remotely.

As you open your latest app to switch on the central heating or warm the house before you get home, remember this: if you can adjust something remotely, so can someone else. Whether it's your boiler, your cat flap or your nursery cam, anything with a power source and an IP address is fair game. If the power source is the national grid, the intention might be to switch on enough devices to overwhelm capacity (see, e.g. Murray, 2018). Peer-to-peer (P2P) devices don't even need an IP address; they can be easily hacked without one. Every day the list of innovative applications adds a new function or device to the list of remote capabilities available to us. The prospect of 'robotic chopping arms' and oven cameras[5] in our kitchens provides a good illustration of the potential domestic dystopia coming our way. While they may offer convenience and a technological high, hooking up these novelties also gifts capability to others who may not have your best interests—or those of your country—at heart.

State activity still stretches to more conventional surveillance and intervention with 'old school' devices as the reporting around the Israeli hacking of pagers and two-way radios used by Hezbollah corroborate (Mekhennet & Warwick, 2024). But technological innovation has made surveillance increasingly intrusive and increasingly hard to detect. If you're lucky enough to be getting a new mobile phone in your stocking this year, malign actors won't need it to masquerade as anything else. Smart phones are startlingly capable of many things, the least imaginative being voice calls and we know that they can be used against us but our behaviour has not reflected the changing threat that the pose. After learning that spyware can be dropped into them almost undetectably and knowing how they have been used to hack journalists, track celebrities and jack bank accounts[6], we still trust our phones implicitly, sometimes unthinkingly. While there's not much we can do at an individual level about state-sponsored activity

[4] Without really understanding how it works—Chapter 1.

[5] https://www.dailymail.co.uk/sciencetech/article-7866553/Kitchens-smarter-CES-tech-not-homes.html.

[6] https://www.theguardian.com/news/2021/jul/18/what-is-pegasus-spyware-and-how-does-it-hack-phones.

like Salt Typhoon and Volt Typhoon (Martin, 2025), we should at least rethink our relationship with our phones, perhaps seeing them more as the loose-lipped office gossip (with unrestricted access to our call record metadata) rather than incorruptible confidant.

As for our own state, our obsession with self-surveillance leaves little need for seasonal disguise in order to see what we're up to since everyone's merrily transmitting it all year round. Without invoking spy craft, our eagerness to share everything has pulled back the curtain on our private worlds. While endless debates about the police using live facial recognition mumble on, the tactic of 'request-and-ingest' for feeds from dash cams has become a standard practice for UK police forces (Sampson, 2026). A rudimentary paste and search in your browser will find hourly police appeals for footage from police forces around the globe. State surveillance appetites are keener than ever to receive data from anything with a lens and we're readily feeding it to them, in most cases with no clear regulatory framework governing what can be done with it. That is not to say this is wrong, just that this aspect of state surveillance is tapping into limitless aggregated capability without a policy word being uttered.[7]

So, what other presents might the AI surveillance ghost of Christmas future have in store? We will almost certainly see more surveillance capability without seeing more cameras. How? Because smaller devices integrated within the fabric of buildings, vehicles and clothing are the evolutionary direction of surveillance. Whether they are a hostile state actor or tawdry amateur snooper, this is a boon for anyone bent on discreet surveillance. For a glimpse into what that future might look like we only need to consider the experience in South Korea. An explosion of tiny devices has created an entirely new species of digital vulnerability to surveillance. *Molka* ('sneaky cameras') built into benign objects like car keys, picture frames and light switches are being used for everything from snaring and sharing political stings to live streaming toilet bowls.[8] Many people have changed their behaviour, even when in the most intimate of settings like their own private bedrooms and avoid staying in holiday rental accommodation or using public bathrooms at all.[9] This, along with the hideous evidence from Mazan, Provence in the Pelicot trial[10] has added a sinister layer of surveillance to already venal

---

[7] For a discussion of the full surveillance implications of this, see Chapter 15.

[8] https://www.scmp.com/week-asia/lifestyle-culture/article/3305971/seoul-palace-installs-spy-cam-detection-public-toilet-combat-illegal-molka-filming.

[9] It breaks my heart. Confronting the traumatic impact of South Korea's spy cam problem on women. https://time.com/6154837/open-shutters-south-korea-spycam-molka/.

[10] The Pelicot trial—a timeline https://www.nytimes.com/2025/10/06/world/europe/pelicot-trial-timeline.html.

criminal offending and the AI-enabled technology of the future is only likely to aggravate it further[11], supplying every definition of the so-called chilling effect of surveillance.[12] Although it may not be next year, when remote biometrics finally meet nanotechnology, privacy as we've understood it will be extinct.[13]

Back to Christmas. Not wanting to be the gizmo Grinch, but each year we make it easier for those with an unhealthy interest in our private lives to intrude and interfere—and that's worth bearing in mind as we bag more smart tech, this season and every other. The UK data regulator recently cited another unseen risk lurking in this area: complacency.[14] Governments and watchdogs issue guidance all year round about giving our data to smart devices but often our surveillance resolutions have the same short half-life as those we make every New Year. AI-powered surveillance realities mean we need to recalibrate the many risks around our homes. We also need to be aware that data which is inaccessible today may be crackable by new technology tomorrow. The Sunday Times has recently warned how China has already harvested vast amounts of data belonging to every British citizen.[15] The Government Communications HQ, which is part of the UK's intelligence service, confirmed that the data had been stockpiled in an 'unrestrained campaign of malicious cyber-activities' by state-sponsored hackers. Covering personal and business messages, commercial know-how and political sensitivities, some of the data extends to information that would be of direct relevance to any attempts to interfere with the National Grid or to spy on people at work but the important thing to note is that the datasets in which the information is contained are secured by encryption keys. Why would securely encrypted data be of any use to another country, or even our own? Because nothing is encrypted for ever. The data has been collated in advance as part of what the news report describes as a 'harvest now/decrypt later' attack and it is believed that hackers are not waiting for quantum computers before grabbing terabytes of secret information from us. The technology is advancing and one day, perhaps not this Christmas or even the next, the data will become unlockable by quantum computing capabilities—and that will bring an immeasurable number of spying gifts for any countries that have amassed encrypted data.

---

[11] https://www.theguardian.com/uk-news/2025/aug/29/chao-xu-pleads-guilty-sexual-offences-greenwich.

[12] See Chapter 18.

[13] For a discussion about privacy and why it is already a difficult concept see Chapter 15.

[14] https://www.cityam.com/uks-data-watchdog-warns-against-complacency-after-fining-interserve-4-4m-over-cybersecurity-breach/.

[15] Dominic Hauschild, Ben Spencer and Diphesh Gadher, China's chilling stolen data plot for everyone in Britain, The Sunday Times 18 October, 2025.

And that's why China—or any other country—might want to hack your fridge.

In the meantime, I've just spotted some lights in the sky. My neighbour has taken delivery of a night vision stealth drone from Shenzhen which is now flying over the rooftops. It's beginning to look a lot like Christmas.

References

Laterza, V. (2021). Human-technology relations in an age of surveillance capitalism—Towards an anthropological theory of the dialectic between analogue and digital humanity. *EtnoAntropologia, 9*(2), 59–74.

Martin, C. (2025, March 20). *Typhoons in cyberspace.* Royal United Services Institute.

Mekhennet, S., & Warwick, J. (2024, October 5). Mossad's pager operation: Inside Israel's penetration of Hezbollah. In *Washington post.*

Murray, S.. (2018, October 17). When fridges attack: Why hackers could target the grid. *Financial Times.*

. Sampson, F., (2026) 'The future of police surveillance and cameras', Chapter 16 in Future Policing, Technology and Transformation Huey, L., Ferguson, L. and Adams, I. (Eds), (2026) Routledge, New York

3 How to Duck AI Fraud? An ABC for Beating Cyber-Scammers

You know the saying: "If it looks like a duck, walks like a duck and quacks, it's probably a duck." In the context of bar room blowhards and armchair professors, this 'common sense' motto may seem pretty harmless and, at times, it can be fun. This 'well, duh' riposte when faced with the patently obvious has become known as the duck test and has been applied as a rhetorical device in recognised fields as diverse as religion, athletics and can even be found in heavyweight scientific papers on serious subjects like cardiovascular surgery (see, e.g. Whiteley, 2019). In the expanding world of identity fraud, however, this commonplace idiom plays a more sinister role. Adopting this laconic approach in the face of potential scams is not only a dangerous tactic, it can also be critical to the scammer's success in defrauding you. And in an AI-driven future this already prevalent threat is likely to become much more problematic.

Before we look at how the so-called duck test applies and why it needs to be beaten, it's worth taking a moment to uncover its origins. The provenance of the test itself is a bit vague. The avian aphorism is attributed to many people, from poets to politicians while several respectable sources trace its more consequential application to American political figures in the post-WW2 period (see, e.g. Immerman, 1982). Wherever it was originally hatched, the duck test was formally adopted and deployed during the McCarthy era as a sure-fire way to spot a communist and its pernicious potential as a process of crude inductive reasoning is well illustrated from that era (see, e.g. Schrecker, 2004). Thankfully we now have more forensically reliable ways of verifying identity, if not political affiliation but the 'I know one when I see one' duckery mindset remains stubbornly nested in many cultures.

F. Sampson, *How AI-Driven Surveillance is Changing our World*, Security Informatics and Law Enforcement, https://doi.org/10.1007/978-3-032-24005-7_3

While fascinating, the application of the test as a tool of political persuasion is not the real point here. The duck test's real power—and therefore its biggest problem—in the context of fraud lies much deeper within us as human beings. As Annie Dillard pointed out in her famous philosophical observation, we see the world not as it is, but as we are (Dillard, 1974). David Robson's *expectation effect* offers a more empirically rigorous explanation of this human trait (Robson, 2022). The potency of expectation within us is extraordinarily potent. Robson shows how, for example, a belief that they have been given an anaesthetic allowed patients to undergo battlefield surgery (the placebo effect) or how an expectation that certain things will be harmful can result in the person having symptoms of the expected harm even though the substance was innocuous (the nocebo effect). More emphatically, his work demonstrates how we can fool ourselves so effectively that placebos still work *even when we know they are placebos.* While the expectation effect is put forward by Robson as a positive phenomenon whereby we can improve our life experience by altering our expectations, in the context of being scammed, our innate tendency to see or feel what we are expecting to see or feel is a significant impediment. The presence of apparently rational belief or the suspension of what we tell ourselves is an 'irrational' disbelief is the fraudster's best friend. It is regularly used against us to defeat even the most sophisticated biometric technology. In other words, because we see what we expect, our—often subconscious—default is to apply the duck test, putting a powerful weapon in the hands of a fraudster.[1]

Scamming tactics vary infinitely and become more elaborate every day. Cyber criminals are routinely using things like homoglyphs (numbers or characters that look identical or imperceptibly similar to authentic ones) and 'typo squatting' (transposing a character into a domain name) to fool us, but the entry-level scamming formula is technically simple, frequently effective and requires no real knowledge or skill on the part of the scammer. The basic recipe involves the same ingredients: present the victim with a dilemma, create a sense of urgency and proffer a solution. For example, you receive an email purporting to be from your bank. The scenario doesn't really matter—what is important is that it requires you to believe something alarming or important like an overpayment on your account, a suspicious transaction on a credit card or a change in the conditions applying to a product. The email will be asking you to do something to your cost—make a payment, disclose some information or simply clink on a link. And you must do it soon so there will be a temporal aspect—a date or deadline. The email looks

[1] When you add convincing digital corroboration such as Deepfakes, the power to deceive becomes even more alarming—see Chapter 14.

like your bank's emails, sounds convincingly like their emails and is followed up by a phone call with someone who knows a lot about you and your account and who makes all the right noises. Even if you've read about unsolicited email and know a bit about scamming, this scammer's tactic appeals to horse sense; it relies on the implied 'obviousness' of its veracity and also on your fear of looking foolish by doubting something when anyone else can see it as plainly genuine. The whole foundational scam relies on your running the duck test—consciously or unconsciously—and deciding that all is in fact as it appears to be.

Deceptively simple, this move is on page one, Chap. 1 of scammer playbooks worldwide. Here's another example. In Southeast Asia there's a current scam that involves the victim being contacted by the police and told they must join a Zoom call immediately. Once connected via the sent link, the victim will see the officers sitting, uniformed and earnest, behind an authentic-looking desk with a plausible police station in the background. The 'officers' deliver an alarming message (a relative in custody, an unpaid historical speeding fine now looking like imprisonment, etc.). Please pay/click/divulge information immediately in you want to avoid imminent law enforcement consequences. Quack!

The forces underlying our deference to the duck test may be emotionally rooted (per Blake's poetic description '*as man is, so he sees*'[2]) or perhaps springs from some atavistic mechanism that has got us humans this far in our evolution. Whatever its origins, if we are to survive in the AI era one practical challenge will be knowing how to beat the duck test. Here the police can help but not in the way you might think—more in the way that *they* think.

Detectives in the UK used to be taught what was called the ABC of investigating crime: **A**ssume nothing, **B**elieve no one, **C**hallenge everything. This traditional approach is founded on inherent scepticism and the application of what later became fashionably labelled evidence-based policing (Sherman, 1998). As an investigative policy it naturally created some significant problems, for example, when it came to interviewing victims who wished to report crimes like sexual offences where complainants felt they were not taken seriously or were suspected of wrongdoing themselves. The intelligence officer's credo of 'trust but verify' is more nuanced but, as a defensive tactic against unsolicited scam messages and unmasking fraudsters, old-fashioned cynicism has a lot going for it.

When applied to unexpected demands or communications, the ABC mindset requires critical thinking. It means checking exactly what you *know* about the message, its author, its timing and urgency, etc., rather than what you have assumed or simply been told. It also means considering whether the consequences

[2] https://findingblake.org.uk/seeing-wood-plant-blindness/.

are really as imminent or dire as they are being sold and it means demanding proof, taking nothing on faith alone. The 'C' (challenge) means first and foremost challenging *yourself*: what does your bank say about this type of contact? Is it normal to be asked to do this? Have they tried to pre-empt any questions you might raise about its authenticity? Is there something—or someone—odd in the backstory they've told you, that needs to be explained away? The ABC makes for a more hardened world but there is a lot at stake from digital fraud. A great deal of misery and suffering is being caused by ID fraud and data hacking, from online gambling to sexual exploitation and people trafficking—there aren't many targets who avoid a scam and later think 'I wish I'd been less cynical'.

Ducking the scam is not as easy as it sounds. Savvy people are left nonplussed because they 'double-checked' just to make sure and, having their suspicions aroused they looked for proof. But, as with the expectation effect, scamming involves a complex interplay of psychological forces. The ground between objectively looking for proof and the phenomenon known as 'confirmation bias' can be very narrow. The latter exists because we naturally tend to see or seek evidence that supports our first theory (isn't that a beak, surely those are feathers?) and give it more weight than information pointing the other way. When you get an unsolicited phone call supposedly from your phone provider, the part of you that stops short of hanging up and moving on is the part most vulnerable to confirmation bias—and most vulnerable to being defrauded. The ABC approach involves learning to distrust the part of yourself that refuses to delete the message immediately, that part that thinks you *might just* have inadvertently run up an overdraft or forgotten to pay a bill or inherited a lot of money. That's the part the scammer is betting on and the odds are usually very much in their favour or they'd stop sending this stuff.

Looking ahead, when AI and deepfakes are involved, the duck test will become many times harder to spot let alone resist. Here's an example.

> At the start of 2024, a finance worker in Hong Kong received an email calendar invitation to join an online meeting by the company's UK-based chief financial officer. While aware that the email might have been a phishing attempt, he joined the call and saw colleagues on the call that he recognised. According to the Hong Kong police, the worker agreed to transfer a total of \$200 million Hong Kong dollars or \$25.6 million. It later transpired that the call was a scam and the 'colleagues' were deepfake video manipulations. Quack!

When you see moving, speaking images of your co-workers on screen, why would you question whether you are in a genuine meeting? How do you resist the instinctive impulse to help if you are in fact hearing the *real voice* of a relative on the phone beseeching you for urgent funds? You know your own kids when you hear them. AI voice fraud and 'vishing' involve taking the real faces, voices and other biometric features of living people and manipulating them. It is already well-advanced and fiendishly convincing. If scammers can persuade someone that they are taking part in a live-streamed meeting with colleagues arranged by their chief finance officer, transferring millions of dollars as a result, they can certainly con you that you're listening to a genuine voicemail from a relative in need of prompt assistance. Adopting the ABC approach will be increasingly difficult in an AI-driven world. In a work environment we place a great deal of trust in our employer's IT and digital security but nothing is unhackable. In the future will people refuse to take part in any meeting or online event until all others 'present' have identified themselves to everyone's satisfaction?

In the domestic context of unsolicited communications like calls or emails, the ABC tactics may mean asking the other person to confirm a pre-agreed safe phrase (whether you've in fact got one or not) or asking for information about something that only you and the other genuine party would understand. This type of initial confirmation may wrong foot a human or an AI, and it could filter some of the more speculative ones (it could also block a few genuine messages too). Most scamming risk however lies in the engagement, entering into a dialogue and subsequent forming of a 'relationship'. Some victims spend hours on a call with fraudsters whom they believe to be helping them sort through the 'problem' when in fact they are simply building the association founded on nothing more than the duck test. I do not have any research data to support the hypothesis, but experience suggests that there is probably a correlation between the length of the interaction with the fraudster and the evaporation of suspicion on the part of the victim.

In the evolution of Large Language Models (LLMs) in AI, one long-standing technical goal has been to pass the Turing test. This is where the technology is so humanlike in its interactions you are unable to tell that you're communicating with a computer.[3] Using the ABC approach may reveal the presence of an AI scam because something looks (or sounds) *too* much like a duck to be a real one, i.e. it's simply too perfect. Going back to Dillard and what she called the 'flawed nature of perfection'[4], by challenging answers from the interaction we might start to detect signals that are too true to be good. In a neat irony, this would mean that running a reverse Turing test reveals that the AI technology's communications are not flawed enough to be one of us.

---

[3] You can try it here: https://turingtest.live/.

[4] *Loc cit.*

In the new world of AI-enabled fraud and identity theft, *anas cybercrimus*[5] is an aggressive and highly venomous species. If an unexpected demand looks like a duck and is waddling and quacking, it's probably not a duck at all but some malign creature hiding behind alarm and urgency, and offering you a way out that you will ultimately regret. Whatever else you do, don't be gulled (!).

References

Dillard, A. (1974). *Pilgrim at tinker creek.* Harper's Magazine Press.

Immerman, R. H. (1982). *The CIA in Guatemala: The foreign policy of intervention* (p. 102). University of Texas Press.

Robson, D. (2022). *The expectation effect: How your mindset can transform your life.* Canongate Books.

Schrecker, E. (2004). McCarthyism: Political repression and the fear of communism. *Social Research, 71*(4).

Sherman, L.L.W. (1998). Evidence-based policing. Police Foundation (US).

Whiteley, M. S. (2019). If it looks like a duck, swims like a duck, and quacks like a duck, then it probably is a duck. What "the duck test" tells us about systematic reviews and meta-analyses of LEED and other EVLA parameters. *European Journal of Vascular and Endovascular Surgery, 58*(2), 243.

[5] Cybercrime duck.

4 Revealed: Why We Can't See the World's Most Surveilled Cities

What was the most surveilled city in the world last year? Beijing? Pyong Yang? Moscow? Perhaps it was Lotsee, Oklahoma (now wouldn't that be fun?).

Our adoption of social media linked by global interconnectivity shows how much we like our ratings and rating our Likes on a minute-by-minute basis. Stories about technological innovation and gadgetry are popular all year round but, as we have seen with the Christmas spy cams[1], there is a growing tendency for a seasonal focus on AI surveillance trends. With that in mind there's another important one that appears regularly in the digital calendar—the ranking of places according to the amount of surveillance they undergo.

As each year draws to its close, the end-of-term summaries and new year lookbacks are filled with bests and worsts, mosts and leasts. It's a time when every headline lists towards league tables: 'the most-downloaded movie', 'the best-selling kettle', 'the least interesting chapter on AI-driven surveillance'.

One such list that is becoming a hardy annual is 'The World's Most Surveilled Cities'. These lists can appear in several guises—Cities With the Most Surveillance Cameras[2], Cities with the Most Government Cameras[3], Spied-on Population, etc.—and they are usually presented in comparative tables, presumably so we can check our own national standing against other regimes and feel duly relieved/appalled. While the spread of surveillance technology has fed our love of taxonomies, that same technology has made any surveillance rankings all but meaningless.

---

[1] Chapter 2.

[2] https://www.comparitech.com/vpn-privacy/the-worlds-most-surveilled-cities/.

[3] https://neomam.com/most-surveilled-citics-worldwidc/.

F. Sampson, *How AI-Driven Surveillance is Changing our World*, Security Informatics and Law Enforcement, https://doi.org/10.1007/978-3-032-24005-7_4

Sometimes—as bloggers and influencers know well—adding a little ordinality to your teaser can drive even more SEO[4] traffic towards your offering: 'the 5 most effective excuses to use at work' or 'the 7 best-kept secrets of vlogging'. There are many lists claiming to reveal the world's most surveilled cities; one year there was even a BBC Top 20[5] which had London coming in at a shocking 3rd place, sitting uncomfortably among other perhaps less liberal countries, with most of the cities reassuringly located in China. As entertainment these rankings can be interesting and even imaginative but as evidence of surveillance activity they are useless.

Despite their tone of forensic certainty, comparative lists of global surveillance aren't proof of very much. Trying to tot up the places on Earth that are subject to the most surveillance activity is calculating the incalculable—you may as well label a city the World's Least Dreamed About as there is no reliable way of measuring either. There are several reasons why this is true.

First is the starting point of the compilers. Most sources proclaiming the relative prevalence of surveillance in a particular location begin with camera counting. This methodology involves looking around the relevant city for overt surveillance cameras, adding them up and then publishing the total. To lend it more arithmetical kudos, you can divide the figure by the official population. But cameras over population figure will not even give you much idea of the number of cameras that are over the population. It can give you mathematical comfort, but the quotient won't begin to tell you how many people were under state surveillance in any period. It might give you an idea of how many *visible cameras could be seen from public spaces* on a given day but that's not even close to being the same as how many times or in what depth that area was observed, still less the people passing in front of them (*you'd do better counting smart doorbells—and there are a lot of those). You could divide the number of visible cameras by the recorded size of the city in* miles/km squared. Do this and you'll have a ratio that you can report in your headline that also sound compelling but, as a measure of surveillance saturation, it's meaningless.

If we take the UK as an example, the majority of street-facing cameras are operated by local authorities, many aren't working or watched and they're not the only, or even principal, source of surveillance data. Counting them—either in person or by digital examination of public records—will not provide much accuracy as to the real number of cameras in place, and none whatsoever about

[4] Search engine optimisation.

[5] https://www.bbc.com/future/article/20201014-totalitarian-world-in-chains-artificial-intelligence.

what those cameras are doing. The more inquisitive taxonomist could approach the police and councils and ask them outright (using statutory powers given to the citizen to demand such information[6]) how many surveillance cameras they have, what functions they run and so on. When I did this as part of my Biometrics and Surveillance Camera Commissioner's national survey in 2023, most didn't know and many didn't answer. This was not particularly surprising given the patchy nature of public space surveillance regulation but, more importantly for our purposes here, any answers that the police and local authorities might have given would have been largely irrelevant as we shall see.

Another barrier to compiling a reliable list is the technology itself. With the exception of a few dinosaur systems, most surveillance cameras have a pan/tilt/zoom (PTZ) 360-degree capability which allows them to 'look' at different geographical spots at differing levels of resolution. They are dynamic and zoom in and out, or pan around either under the direction of an operator or, more often now, in automated response to a detected movement or programmed instruction. Even basic surveillance cameras come with built-in capability beyond just capturing images, with features such as sound recording and automated number plate reading (ANPR) as standard. Staring up at one from the street will reveal nothing about which functions a camera is operating at that moment or how many people are accessing its data and from where. As a result, it is very difficult to know what a camera has been looking at, listening to, counting or comparing. This is an elemental problem even with spatial surveillance. Watching people rather than places makes up for a far greater proportion of digital surveillance activity, and we have not even started to consider more advanced biometric capabilities such as facial recognition and matching.[7]

If all the police and councils diligently and accurately collected and reported all the performance data of all their surveillance cameras, including body worn and aerial borne, we still would not be much closer to having a reliable number. This is partly because an increasing amount of 'public' space where people gather, meet, work travel and shop is commercial space, privately owned and commercially monitored. The surveillance systems are also privately owned, operated and monitored, and even though we can see many of them when we visit those areas, we do not know how many or what types of camera they are operating over what period of time or what spatial area how many of the images have been shared with the police. We would also have no information at all about

[6] Freedom of Information Act 2000 https://www.gov.uk/make-a-freedom-of-information-request.

[7] Discussed in Chapter 11.

the many other cameras in police vehicles, body-worn video, drones or other remotely operated systems.

So, camera counting might just give you a rough estimation of *visible and working cameras* covering *some (unknown) area of public space* but, even if that were possible, the figure would be of no real help in understanding the extent of surveillance in a street let alone a city or country.

But perhaps the main reason why using the mathematics of visible camera infrastructure is a flawed method of assessing citizen surveillance is us—the citizen. Driven by techno-convenience, we are far busier than even the most intrusive state surveillance organs. We are feverishly recording everywhere and uploading images of everything as part of our daily lives and routinely share the product with the police.[8] Internet scrapers might have some idea of how many selfies—and images of passers-by—were uploaded at, say, the Trevi fountain, ESB, Machu Pichu or Sydney Opera House last year and of course there are lists of 'the World's Most Photographed Tourist Attractions' (most of them different and similarly unreliable). But how would we calibrate all this snapping and streaming and sharing and factor it into our rankings?

To figure out who was subject to the most surveillance and where, you'd have to take account of features such as the multi-modal recording devices in hundreds of thousands of vehicles wending their way through our cities, but no one knows how much data they capture and hold about people and places. As I have noted elsewhere, land, water and airborne vehicles contain multiple recording devices capturing data as they navigate in, around, above and below our cities but the amount of data they hold about people and places is incapable of being counted. As an aside, once those vehicles become self-driving, they will need automated total data capture of their surroundings in order to function and our ability to count what they see will move even further away from us.

We must also take account of the fact that vast amounts of images are of no immediate practical use as surveillance material unless and until they are later compared with known faces and other pictures. Images recorded and retained in perpetuity have a latent surveillance impact that may not be realised for years. Do we count the capture of this not-yet-surveillance data in our calculations?[9]

---

[8] This has completely changed the 'surveillance relationship' between the citizen and the state—as discussed in Chapter 18.

[9] This is one reason why we should care about AI-enabled retrospective image analysis, something we look at more closely in Chapter 6.

Then there are the strategic level surveillance tools like official national networks counting mast 'pings' of mobile phones[10] and ANPR records of vehicle identification and movement. In the UK, policing alone claims that its ANPR system has more hits-per-second than Instagram and is approaching more than 90, 000, 000 reads per day. While impressively voluminous, the accuracy of that data is itself dubious, putting on another list: the world's least reliable ID systems.[11]

Combining social media analytics with other mass metrics like call record metadata, biometric-based financial services, mobile driver's licences and digital ID wallets would give us a less wrong guesstimate than visible camera infrastructure but, when compiling world surveillance rankings, we'd still be light years from podium territory.

All of these surveillance realities present insuperable difficulties to identifying the most surveilled people or places on Earth but possibly the biggest technical hurdle for claiming the title of 'most-watched' is this: surveillance doesn't need to be a) located in, or b) visible from a city in order to look at it and its inhabitants. If you're unsure about this ask the military. Satellite technology means that pretty much everywhere can be watched from somewhere. This is not just bad news for ranking fans but also for those who enjoy quoting that other surveillance list favourite 'Things Visible from Space' (which now includes virtually everything[12]).

Satellite technology and AI-enabled surveillance capability make comparative calculation all but impossible. The real revelation by those relying on crude camera counting is that they simply have no idea about surveillance, even if we confine that term to spatial visibility. The available technology means that anyone with the right hook up can look around the world like a scientist moving a slide around under a microscope. Global surveillance capability defies traditional methods of comparison and conclusion as the following example shows:

> In the province of Soria, Northern Spain lies a hamlet called Tajueco. In the rankings of cameras per head of population or kilometres squared Tajueco is unlikely to make the top 100, 000.[13] In 2023 a Cuban man living in Spain was reported missing by his relatives who were concerned about his sudden disappearance. Fearing for his safety the police began an investigation in the area where he had been living. Several months

[10] See Chapter 15.

[11] To see why, skip to Chapter 13.

[12] The variant, 'things visible to the naked eye from space' is even worse as there is currently no possibility of a technologically unassisted 'naked' human eye being in 'space'.

[13] Google cites the population as 56-57 people and the number of 'dedicated surveillance cameras as zero.

> into the inquiry officers looked at Google Maps and saw that a passing mapping car had recorded an image of someone bundling what might have been a body into the back of a red car parked in a quiet side street in Tajueco. The police subsequently acknowledged that the image had made a significant contribution to what became a murder investigation which led to the arrest of two suspects.[14]

This story serves two purposes. First, it illustrates how the visibility of our world is expanding without any need for visible street cameras. Second it underlines the challenge for the effective regulation of surveillance. How do we begin to assess the *surveillance reach* of an upload or advertisement that's viewable by anyone on Earth with access to a router from anywhere on Earth? How do we compare the *surveillance impact* of a single Google Maps camera image from a town with a population less than 60 with, say, 10, 000 hours' footage of a small patch of pavement from a static CCTV camera in Tokyo (pop 37 million)? The answer is simple and it's the same as the answer to the question 'how do we know the most surveilled city on Earth'? We don't.[15]

Uncounted and uncountable billions of images and other surveillance reference points are being or will be, captured and shared across thousands of cities, this year, next year and every year to come. When all this data needed a human to join the dots it wasn't much use; no one was going to live long enough to get through a fraction of it. Artificial Intelligence has transformed all that, alchemizing what was digital chaff into surveillance gold. Privacy advocates habitually reach for Orwell's fiction when highlighting surveillance risks; the reality is more like Bentham's (1995) panopticon, except the 'building' is the world and we, the population, are its interns.

Apart from corroborating our love of lists, ranking the 'most surveilled' cities illustrates one thing: a misunderstanding of our technologically interconnected lives. Pinpointing areas or people that were surveilled (that seems to be a verb as well as an adjective) has so many variables and we can't access the data

---

https://www.google.com/search?sca_esv=c289d407d4feae92&rlz=1C5OZZY_en&q=%22Tajueco%22+surveillance+cameras+per+head+of+population&sa=X&ved=2ahUKEwjy_oi-rtWQAxWPa0EAHR0-GIsQ5t4CegQIGhAB&biw=1440&bih=703&dpr=2.

[14] https://www.theguardian.com/world/2024/dec/18/google-maps-car-snaps-vital-clue-in-spanish-missing-person-case.

[15] In another example of 'private space', someone browsing a property website noticed a picture on the wall of an Argentinian home that was being advertised for sale. Subsequent enquiry revealed the painting to be one of many missing works stolen from a Jewish art dealer in the Netherlands during the Second World War—https://www.theguardian.com/world/2025/sep/04/nazi-looted-old-master-painting-giuseppe-ghislandi-recovered-argentina-property-listing-ntwnfb.

needed to compare them. Rating them does nothing to promote understanding of surveillance in the AI era, and even less to further its regulation. A more useful list would be one containing the *standards* by which the state will be held to account for its use of any AI-enabled surveillance capability—from satellites to dashcams. At CENTRIC we are working with citizens and law enforcement in many different countries to compile it.[16] Were we to make one, our next New Year's surveillance resolution would be to shift the focus from countability to accountability, to find which cities have the greatest level of surveillance awareness or the best AI transparency ratings for their police. Nominations for every year are now open.

Reference

Bentham, J. (1995). *The panopticon writings* (pp. 1748–1832). Verso.

[16] https://aipas.co.uk/.

5 Why CCTV Has No Future (and Even Its Past Is in Doubt)

How do you prove why someone didn't commit a crime? Measuring *reductions* in crime is easy enough—just show a comparative fall in reported/recorded offences over a chosen period. But proving the contribution that your technology made to that reduction is notoriously difficult as it involves trying to show how much of something that didn't happen didn't happen because of what you did. This has been the enduring public space surveillance challenge for closed-circuit television (CCTV) over the past 50 years and the response hasn't been very convincing.

Given the alacrity with which towns and cities and governments have continued to buy up and install CCTV camera systems, one would expect there to be a substantial body of evidence demonstrating that they prevent crime and criminality. But one would be wrong. Notwithstanding the considerable amounts of public money that have been sunk into CCTV projects across many countries over several decades, the research into their impact on crime has been largely inconclusive (see, e.g. Phillips, 1999). However, in the AI-enabled surveillance future all the research of the past may not matter.

Asked to think of 'surveillance', most people will picture a classic CCTV system. Cameras—usually white and invariably positioned above head height—look down on somewhere and recording any 'someones' passing within range. In this typical cinematic scenario, the images are relayed back to a central control room facility where people monitor banks of screens, intervening and directing as the plot dictates. This is very much the default model used in both political and fictional discourse and usually references the worrying 'Orwellian' future ahead of us. In this version of the future, the physical surveillance infrastructure of the state sits front and centre, becoming increasingly prominent and inescapable as we are impelled towards technological dystopia.

F. Sampson, *How AI-Driven Surveillance is Changing our World*, Security Informatics and Law Enforcement, https://doi.org/10.1007/978-3-032-24005-7_5

There are still many traditional CCTV systems in our streets and on buildings, and I am still invited to speak at 'CCTV conferences' attended by national 'CCTV groups'.[1] However, the reality of surveillance ecology is nothing like this and we have seen how the present shape of state surveillance has followed a very different path.[2] A lot of public and private space surveillance is now conducted by multi-modal devices, accessed remotely by anyone with the permissions and connectivity to do so, from anywhere in the world. The AI-driven surveillance model will replace traditional CCTV that we have grown up with, and the focus of surveillance will be people rather than places. The innovative systems beginning to watch us are not on closed circuits and have nothing to do with TV. This is not just labelling—the technological and societal differences from CCTV are elemental. AI-driven surveillance systems are a completely different species from CCTV and, as we will see, there are several reasons why the distinction between 'old school' perceptions of CCTV and the realities of AI-enabled surveillance are important to our future. To understand why, we need to take CCTV apart.

When we consider how traditional CCTV worked, we can immediately see its in-built limitations. The 'surveillance product' was pre-determined by a number of inherent features. Attaching static cameras to fixed points like lamp posts, gantries and buildings restricts their view to the angle and aspect at which they are set. Although multiple cameras could be attached to the same support and pointed in different directions, the combined picture was still incomplete and immutable.

Surveillance is the art of perspective and that affects what any of us can see for ourselves. The original static CCTV camera's perspective was limited, both by where it was sited and what it was sighted on. Reviews of legacy CCTV systems often discover old cameras looking dutifully at a piece of ground beneath it that used to be a car park, watching over cars that are no longer there (raising a slightly philosophical question: if a tree falls in a forest with an unmonitored CCTV camera, at what point is it under surveillance?). At the end of the Cold War there was a healthy market for second-hand CCTV cameras that had been stared fixedly at the same grey street corner or dark doorway for so long that a ghostly image was forever burned onto their lens. These watermark images appeared as a kind of surveillance palimpsest in any future recordings that the liberated cameras made in their new jurisdictions.

But it's only by going *behind* the lens and reflecting on what was happening inside the traditional static CCTV camera that the real constraints appear. Even

---

[1] https://www.publicpolicyexchange.co.uk/event.php?eventUID=PH05-PPE.

[2] See Chapter 18.

with the arrival of dynamic variants using pan, tilt and zoom (PTZ) functionality and streaming live images to control centres, CCTV cameras were primarily ingesters and regurgitators of images and sounds.[3] From its limited vantage point, the CCTV camera was often recording within its limited field of vision onto a magnetic tape or disc to be collected and compared later by someone with the authority and motivation to do so (picture the 'TV cop asking a night club owner for security tapes conveniently wiped after the murder' cliché). If all those surveillance images and sounds were to be of any use for evidential or intelligence purposes, they had to be viewed, collated, edited, sorted and compared—by a human. Increasing the number or coverage of cameras simply increased the volume of material that would have to be combed through and combined by someone. With CCTV, having more cameras without more people to undertake the processing doesn't mean an increase in surveillance impact; it just means more cameras[4]—unless you can demonstrate that the presence of cameras alone deters crime. And as we come to the end of the CCTV era, we can say that the presence of cameras on some types of crime has been emphatically and empirically *disproved*. For proof, we need only look along any British high street.

> In 2025, shop thefts in the UK reached a record figure of over 50,000 per day.

> According to the Economist[5] shoplifting octupled in a decade and amounted to around 20 million thefts annually. Fare evasion on London's Tube and bus network reportedly doubled since 2019 and opportunist street crime made it the phone snatchers' capital. There are various theories as to why this happened but consider this: whether they're supermarkets or corner shops, pop-up vendors or department stores, airport lounges or filling station forecourts, every one of the retail venues behind those statistics will almost certainly have some form of CCTV security cameras in place. With over 70, 000 public-facing cameras, London Underground has one of the most concentrated CCTV surveillance systems in the country if not the world, while the square footage of our streets and bus stations and parks covered by CCTV has never been higher.[6]

In the summer of that year, the UK government announced a 'one in one out policy' to herald the new Anglo-French strategy for tackling the immigration crisis.

---

[3] For a discussion of the eras of surveillance devices, see Chapters 1 and 14.

[4] But in the future we may see *fewer* state surveillance cameras—see Chapter 15.

[5] https://x.com/TheEconomist/status/1944673227705364731.

[6] For a discussion on why that is a pointless metric that doesn't help us understand the surveillance impact, see Chapter 4.

It was also an apt description of how—despite having extensive CCTV coverage—some London stores were responding to rampant retail crime by locking their doors, only allowing one individual shopper to enter when another left.[7] The official figures published by the British Retail Consortium meant that the number of shop staff assaulted at work would fill London's O2 Arena every 10 days, while the annual cost of theft passed £2.2 billion.

The numbers were stark but the basic challenge for retailers can be easily summed up: how to stop unwanted people walking in the door and £millions walking out. On any view, the evidence from retail and public transport indicates that the answer wasn't more CCTV cameras.

If the purpose of installing CCTV was to enhance security, prevent crime and provide evidence for the prosecution of offenders, it's not really going out on a limb to say that those cameras haven't been working. Despite signage advertising their ubiquitous presence, CCTV has failed to deter criminality in these settings. Why?

One answer may be CCTV's inherent limitations that we considered above. It is those limitations that drive the risk and consequences of being caught on camera. When I walked through a British city in 2015, the likelihood of my being identified let alone located simply by coming within range of one of its many closed systems of fixed-point street cameras, having my image extrapolated, examined and edited by someone able to match it with other images from other systems was low to non-existent—and if I had reason to avoid being caught on camera, I probably knew this. The voluminous research into CCTV may be inconclusive but it's also going to be irrelevant to the future of public space surveillance for two main reasons: the nature of the technology and the knowledge of the citizen.

To see why, we should look at them both: the technology and us, the citizen. Surveillance devices are becoming so much more than 'cameras'. Many intelligent security devices are networked computers. Those with facial recognition functionality aren't just scanning a limited space and recording whoever passes within their 360-degree, long distance field of visibility; they are actively looking for people and they 'know' who those people are. Fully programmed to find me, the AI-enabled surveillance system will not only know what I look like, but also what I sound like, how I walk, what I wear and where I've driven—not just 'on the night in question' but generally. In the future it will have access

[7] https://www.lbc.co.uk/news/uk/ocked-doors-security-crackdowns-shoplifting/.

to biometric data like my heartbeat (which is as unique as my fingerprint[8]) and will be able to detect them through walls and structures.[9] The smart surveillance device will know where I have been, where I am probably going, when and with whom—again, not just at the time in question but habitually. It can already teach other devices and learn from them beyond its original programming; it can retrieve, combine, recall, synthesise and share information in huge volumes and at extraordinary speeds. It can communicate and problem solve. And every time it looks for me, the smart surveillance device and its partner systems get better at finding everyone.[10] These features are crucial differences from, and incomparable with, the closed-circuit camera.

As for the citizen, once I know devices capable of recognising me instantly by matching my features to libraries of images like passports and driver's licences are actively trying to find me, I am walking a very different road from that overseen by the benign CCTV street camera —and I don't need to be a surveillance expert to see it.[11] These sophisticated surveillance tools were once the preserve of state intelligence operatives and if I am unsure how they can be used—and misused—world class know-how and 'as a service' technical support are only a mouse click away.

It is a reasonable hypothesis that this combination of capability + awareness will have some impact on offending in a way that CCTV could not have been expected to do. Of course, that hypothesis will have to be tested but we probably won't need to wait another 50 years: it has already found substantial support from the early deployment of live facial recognition (LFR) in shops.

Having finally realised that their cameras weren't working, some retailers turned to biometric solutions. Stores deploying LFR reported reductions of up to 70% in crime—not just at the start but on a sustained basis. Early use of the technology is clearly helping retailers assert some control of their business space in a way the CCTV has consistently failed to do. As to why this might be happening, the results will need to be analysed on a longer-term basis but one potential reason for the dramatic early impact may be the way in which in-store crime happens.

---

[8] https://www.biometricupdate.com/202203/human-heartbeat-can-be-accurate-biometric-identifier-study-concludes.

[9] https://ddrc.uk/wifi-routers-can-see-people-through-walls-heres-how-it-works/.

[10] Meaning it will soon be almost impossible to avoid being found: see Chapter 16.

[11] In Chapter 1 we consider how we are all using AI-driven surveillance technology without needing technical knowledge.

The most effective way to stop people doing unwanted things in your shop is to withdraw their invitation to enter. This was the principle behind the 'one-in-one-out' reaction above but that was an unsustainable strategy for any major retailer. With CCTV cameras, it is very difficult and labour-intensive to spot the few people who should be excluded among genuine customers. LFR however is perpetrator focused, identifying individuals who have previously caused harm or loss and alerting staff who can deny entry, quickly and confidently. In the retail context LFR matches the solution to the problem. This 'identify and deny' approach is not going to suit all crime types in all settings[12], but it is a good example of how biometric surveillance prevents some crime from happening in the first place. The early impact of LFR on areas like retail crime is already measurable, verifiable fact, while after 50 plus years and millions in public money, the same still cannot be said of CCTV.

In such an AI-enabled surveillance environment there's something quaintly British about the fact that we still talk about 'CCTV' but there's nothing endearing about basing the regulatory framework for public space surveillance on the CCTV model[13] and the Orwellian vision. Both are historical surveillance artefacts. AI-driven surveillance did not evolve out of the monolithic infrastructure of cameras on sticks and the images they captured. CCTV is photography, and LFR is biometrics—two very different disciplines processing data in very different ways.[14] This is one reason why trying to 'convert' CCTV cameras into advanced AI-driven biometric systems as one London Borough has chosen to do[15] is highly problematic and risks creating a Frankenstein effect in public space surveillance. Running facial matching capabilities in a street is a different surveillance activity from CCTV image capture, technologically, legally and societally. A new surveillance infrastructure of integrated biometric processors masquerading as CCTV cameras will be fundamentally flawed. If the state is to be publicly accountable[16] for its use of AI-enabled surveillance capabilities, it will need a very different approach from that of the CCTV generation.

---

[12] Nor is it a smart way of alerting the police—see Chapter 17.

[13] https://www.gov.uk/government/publications/update-to-surveillance-camera-code.

[14] The UK's Data Protection laws treat biometric data as special category personal data requiring much higher levels of care and accountability—https://ico.org.uk/for-organisations/uk-gdpr-guidance-and-resources/lawful-basis/special-category-data/what-is-special-category-data/.

[15] https://www.biometricupdate.com/202509/as-london-expands-cctv-network-critics-express-concerns-of-overreach.

[16] What accountability means and why it is necessary is explored in Chapter 10.

Reference

Phillips, C. (1999). A review of CCTV evaluations: Crime reduction effects and attitudes towards its use. *Crime Prevention Studies, 10*(1), 123.

6 Do the Police Ever Forget a Face?

There is a neurological condition called prosopagnosia which the National Center for Biotechnology Information defines[1] as 'the inability to recognize known and new faces'. Some published studies of the condition—also known as face blindness—describe symptoms as including impairment of 'facial recognition', 'face recognition', 'face identification' and the 'forgetting/not remembering of faces' (Josephs and Josephs, 2024).

Is there a difference between face recognition and facial identification, between forgetting and not remembering a face? In medical terms perhaps not but in biometric terms—particularly in law enforcement—there is, and it can be profound.

Many cameras can recognise a human face as distinct from, say, a logo or a book. Next time you use your smart phone to take a picture look out for those yellow (if you're using an iPhone) squares that instantly form around any faces in view. Where it puts a frame around a face, this bounding box function is also known as face detection. Face detection comes as standard in most smart device cameras and in many you can switch it off. This basic ability to distinguish a face from any other objects within the camera's view could be described as literal face recognition and the camera as a face recognition camera: it knows a face when it sees one. An entry-level capability to pick out what may be a human face will be of limited use in a law enforcement context. There are use cases where it might help, for example, if the police are searching or scanning an area or sifting through images captured by other cameras for signs of human presence.

[1] https://www.ncbi.nlm.nih.gov/books/NBK559324/.

F. Sampson, *How AI-Driven Surveillance is Changing our World*, Security Informatics and Law Enforcement, https://doi.org/10.1007/978-3-032-24005-7_6

But generally speaking, being able to identify what appears to be a human face is not the same as being able *to identify the person whose face it might be.*

Therefore, while in the context of a neurological study, the brain's capacity for face identification may be similar to face detection, and in biometric surveillance it's very different.

Before it can 'recognise' who you are, a camera needs to have something more. To start with, it needs something specific and precise to compare your detected face with. The process of making this comparison is better described as face *matching*. Face matching is what's being done when your device wants to confirm your identity before giving you access to your personal account with an online service provider or to unlock your phone. The technology isn't saying '*this is person A*', it's saying '*this face is a match with the image that has been liked with this device or account*'. Face matching in this scenario is a simple 1:1 process where the device matches your single image with the one planted in its 'memory'.

Advanced technology means that face matching can now also be undertaken on a 1-to-many (1:*n*) basis. Examples of how this has been used include criminal investigations where the police have seized recordings in child sex abuse investigations and want to compare a picture with images of victims and offenders. While this may produce a match, there may still be no information as to the specific *identity* of the person. This form of 1-to-many face matching can give you a confidence assessment that the person in video X is the same person as the one caught on dashcam B or security camera Z. With this type of face matching, there is 'recognition' of the image being a human face and no risk of 'forgetting' it: if it's on the database, it's remembered and can be recalled.

Using facial technology to confirm someone's *identity*—i.e. to 'recognise' them in a human sense—requires a further step involving pre-verification of the library or probe image. In this case, the technology is confirming that '*this face is a match with the image that has been confirmed as being person Y*'. The police already use this form of face matching extensively to confirm the identity of a person in front of them, for example where they want to establish that the driver in the vehicle they have stopped is in fact the person the driver is claiming to be. In this scenario the live capture image of the driver's face is matched against a verified image stored image within a record somewhere that is accessible by the officer, perhaps on a department of transport database of licenced drivers or a police database of people who have been convicted of criminal offences. The most common state use of biometric 1:1 face matching is probably at passport and border control points. The ability to access and match live images with identity databases is a key component in the reach and consequences of state surveillance as we shall see.

Moving up another level is the process whereby an AI-driven algorithm behind a camera detects all the faces of people passing by and compares them instantaneously against a lineup of images with which it has been programmed. This is watching and matching. It is a dynamic form of facial recognition, comparison, matching and identification and is what people often have in mind when talking about Facial Recognition Technology (FRT). In this instance of 'live' FRT the algorithm is actively watching and instantly matching images it captures of passers-by with the images of faces that have been shown to it. In a law enforcement context this works a bit like a police dog—wave an item of clothing under its nose and goes off in search of anything matching that scent.

Unlike the police dog, the algorithm will be programmed with a minimum level of congruity between the face in the street and the image in its memory before registering a 'match'. In other words, it needs to detect a given level of similarity between the face in front of it and the image it's being programmed with. In practice, if you are walking past the camera and the algorithm does not detect enough of a match between your passing face and the image in its gallery, your picture is not kept and the surveillance system has no further information about it or you. This is important because, although the camera has 'recognised' your face as being a human face, it can't identify you—now or in the future: it never 'knew' who you are. It cannot 'forget' what you look like for the same reason. This is a form of digital prosopagnosia and is designed into Live FRT cameras currently used in UK policing. As such, live FRT is arguably far less intrusive than constable snapping pictures of people in the street and keeping the prints for decades. It is also less intrusive than traditional CCTV systems which keep vast numbers of images of everyone that passed beneath their many cameras.[2]

Not all police use of FRT is done by 'live' monitoring and matching. By far the greatest number of images used for FRT matching by police in the UK has been what is known as 'retrospective' application. In 2025, the figure from Freedom of Information request responses by 31 of the UK's 45 police forces was 150 million.[3]

While live FRT cameras operate overtly carrying out real time matching, retrospective face recognition is very different. Taking place after the event—perhaps

---

[2] To see why CCTV has no future in an AI-enabled surveillance world, see Chapter 5.

[3] https://idtechwire.com/uk-police-accessed-150-million-civilian-images-for-facial-recognition-advocacy-group-says/.

many years after—retrospective matching involves taking images of faces from crime scenes or events and comparing them against available images of known individuals. The retrospective facial matching can be done using an image of a face captured by any device ever: old CCTV footage, new smart doorbells, company publicity shots or employee parties, military passing out parades or passers-by captured on dash cams and body-worn videos, funerals and weddings, or oceans of faces at sporting, musical or political events.

AI-driven surveillance capability is making all our available images accessible to the state. If they can search against old family photographs to solve cold case murders or use transport system CCTV captures to find missing vulnerable people, why wouldn't the police do it? What would stop them? Until now the answer has been resources. When it needed a human to do it, individual image matching was a cost–benefit decision, weighing input against likely outcome. In the future, AI-enabled technology will make mass matching so cheap and easy that all the risk—to both the police and the public—will lie in *not* using it in case they miss something.

As we have seen, a critical feature of the FRT systems used in police watching and matching is the comparator or probe image—the lineup photo against which people's faces are being compared. The range of databases of verified images against which the police can match live faces will determine the extent of the surveillance.[4] The greater the range of verified images available to the police FRT systems, the more intrusive their surveillance capability will be. When considering the impact of the police using FRT in your street or railway station, one of the most important questions is how your face might get into any database that the police matching system can look at. This issue is one of the most important policy considerations governments are planning to use FRT, but it is often overlooked (Sampson, 2025). And the compilation of match or watch lists is not always straightforward.

The most obvious list for the police to use for facial matches should be their own. All police forces have databases of images of people convicted or suspected of committing crimes. It's not only reasonable for them to match against these databases when trying to identify some suspects or wanted individuals passing their live FRT cameras, it's essential. But does that mean that the police should be allowed to search against every image they hold of anyone? How about someone who was arrested but then released with the status of no further action (NFA)? There are several million people such people in England and Wales, people who were photographed by the police but never charged with anything. While the

[4] Unlike camera counting which tells you nothing—see Chapter 4.

police undoubtedly have the power to take photographs (and fingerprints and DNA samples) of people in custody, can they retain those pictures indefinitely once the person has been released NFA? Not according to the High Court which ruled in 2012[5] that the police must delete those images and the accompanying data. Despite the court's ruling that the deletion must happen within months of the judgement, by 2025 many millions remain on the database which is managed by the government. Many (most?) of those faces and personal identification data are still in the police memory bank largely because the database has no bulk delete capability.[6] If those individuals want their face to be 'forgotten' by the police, the person must know that they are on the database and apply for their image to be deleted.

When those images were accessible for routine police work, their continued unlawful retention was problematic, but when researchers found that some of the people had found themselves on a police live FRT 'watchlist' without having any criminal convictions (Radiya-Dixit, 2022) the issue became very significant. This was no longer just a data processing compliance issue but something that undermined trust and confidence in the police use of AI-enabled surveillance (Ryder, 2022).

Whether or not the police should be able to hold onto custody photographs for ever, mugshots are not the only way that your picture might end up being matched. Databases of licenced drivers are necessary for the enforcement of road safety and road traffic regulation but allowing unlimited access by police AI tools for intelligence purposes has raised some concern.[7] Similarly, passport photographs must be searched against for proving identity at borders but does that mean that all photographs that we must give to the state can be accessed by police AI surveillance tools for uses unconnected with their principal purpose such as driving or international travel?

The greatest challenge for setting limits on the extent of watching and matching by police AI surveillance capabilities in the future will come from the largest source of facial images on Earth: us. Uploaded selfies and social media posts contain billions of faces, sharing our memorable (AKA matchable) moments in a way that has yet to sink in fully within the surveillance context. The explosion in citizen-generated data is fundamentally changing the surveillance relationship,

---

[5] *RMC & FJ v Commissioner of Police of the Metropolis and Secretary of State for the Home Office* [2010] EWHC 1681 Admin.

[6] Which is no legal defence [Catt] and is a design fault for which they are accountable.

[7] For example by being linked to new powers in the UK's Crime and Policing Bill https://www.biometricupdate.com/202502/uk-revisits-proposal-to-make-a-vast-police-database-from-driving-licenses.

between us and the state and also between ourselves.[8] In the context of retrospective facial recognition there is no need for a police watchlist—there's a universe of data within which AI-enabled tools can look, match and potentially identify us. Retrospective facial recognition is undefined, unseen and its expansion is unchecked.

We have yet to answer the question of how far the state should be able to match wanted people against images freely and frequently shared across social media platforms. The seminal EU AI Act[9] banned internet scraping but why shouldn't the police cast their investigative eye across a virtual crowd to see if their algorithm recognises anyone, particularly if that would save lives or safeguard national security?

As AI-enabled (AI-driven, AI-equipped, AI-enhanced[10]) face matching capabilities evolve, such questions will become more important. Surveillance algorithms of the future won't just be matching against images but inferences. Using genealogical information, they will predict what you will look like before there's a you to recognise and zero-shot learning means that they *will* be able to 'recognise' faces they have never seen before. What will we call that? It's not neuroscience but, when it comes to facial *cognition*, we are going to need some new labels.

The UK's regulation of intrusive surveillance distinguishes between overt and covert with the arrangements for the latter being gold standard.[11] In contrast, retrospective recognition application is invisible to the citizen and mass matching takes place without any formal accountability framework which seems wholly at odds with the rationale for accountability in the state's use of covert surveillance.

The use case for using facial recognition to identify suspects and solve historical crimes is compelling but recent concern at how UK policing has become involved in 'non-criminal' enforcement activity[12] raises questions about where the boundaries should be. How will the police use mass facial matching for intelligence gathering in relation to public protests, industrial disputes or pandemics[13]?

---

[8] See Chapter 18.

[9] https://artificialintelligenceact.eu/the-act/.

[10] https://www.biometricupdate.com/202501/biometrics-institute-identifies-dire-need-for-clear-language-in-biometrics-and-ai.

[11] https://www.gov.uk/government/publications/covert-surveillance-and-covert-human-intelligence-sources-codes-of-practice/covert-surveillance-and-property-interference-code-of-practice-accessible.

[12] https://www.bbc.co.uk/news/articles/cwyp1gk0n23o; bbc.com/news/articles/c0kn54vj55xo.

[13] See Chapter 19.

Policing is about connecting the dots that is what facial recognition does so well, looking Janus-like into the past and the future, making it a formidable—and arguably indispensable—investigative tool. How this will fit into existing accountability and regulatory frameworks has yet to be established but the College of Policing's authorised professional practice for England and Wales says that one legitimate use of retrospective facial recognition is to find 'potential witnesses'. Such a definition could be applied to any living individual in the world and, by that measure, the law enforcement reach of retrospective identification is illimitable.

Does all this mean that, in the future, the AI-empowered police will never forget a face? That remains to be seen, but technology is making it increasingly unlikely that they can't remember one if they need to.

References

Josephs, K. A., & Josephs, K. A., Jr. (2024). Prosopagnosia: Face blindness and its association with neurological disorders. *Brain Communications, 6*(1).

Radiya-Dixit, E., (2022) A sociotechnical audit: Assessing police use of facial recognition. Minderoo Centre for Technology and Democracy, University of Cambridge.

Ryder, M. (2022). *Independent legal review of the governance of biometric data in England and Wales*. Ada Lovelace Institute.

Sampson, F. (2025, April 2) Why we should care about retrospective facial recognition. *Daily Telegraph*.

7 Convenience or Security—Which Comes First?

The use of surveillance by governments is traditionally characterised as a privacy vs security horse trade. The trading generally goes like this: safeguards are intrusive; if we want the state to keep us safe, we must accept that it needs to see what's going on. To be safe you have to surrender some of your privacy. That necessarily involves incursion by law enforcement bodies into our private lives.

In this classical argument, privacy and security lie on a continuum, and if you want absolute privacy,[1] the state cannot protect you.

Of course, the reality is much more nuanced than this. There isn't really a lot of 'trading' involved in state surveillance—it's not a consensual activity from which I can opt out,[2] it's not contractual, and there's not much room for individual bargaining or standing your ground. The boundaries are much more convoluted than a straight-line scale and someone must decide where those wavy lines are drawn. Line drawing for this purpose is primarily a legislative matter for each country (or state in a federal jurisdiction), reinforced by guidance from regulators and reflected in policy by law enforcement agencies. Ideally, the boundaries encompass the technology, the law and the expectations of the citizen.[3] With clear lines in place, the extent of both intrusion and surrender will be accessible to the citizen and managed by courts and regulatory bodies. Without them, the picture is much more blurred.

An example of how regulation *should* work in terms of limitations and considerations is the traditional street CCTV camera. The decision to procure and install

---

[1] A technical impossibility—see Chap. 15.

[2] Even in the setting of live facial recognition vans used by the police—see Chap. 6.

[3] The surveillance triptych running throughout this book.

F. Sampson, *How AI-Driven Surveillance is Changing our World*, Security Informatics and Law Enforcement, https://doi.org/10.1007/978-3-032-24005-7_7

a camera system will usually be taken by a local authority following a consultation with residents, businesses and other partners like the police, and the system will be governed by published policy. However, with the technology changing rapidly and our private use of it expanding in ways that were unpredicted by the preceding generation, hard lines are getting harder to draw. Privacy was already a tricky concept to pin down before AI-enabled surveillance,[4] while the breadth of 'security' is defined by the state, giving its boundaries an infinite elasticity. For a better example of the direction in which things are really going, we should head to the airport.

International air travel requires a high degree of pre-planning and compliance with strict formal regulations. It entails registration with authorities, essential requirements around proof of identity, plus entitlements to leave/enter the countries involved and, of course, security. Satisfying these state requirements is both critical to the journey and quite inconvenient, with all its queueing, and document production, verification and validation. Here the availability of biometric technology to support identity authentication is making international air travel considerably easier for all concerned.

If we replace the somewhat baggy idea of privacy with convenience, we can get a better picture of how things are changing in the world of public surveillance. The financial services sector has led the way in bringing biometric technology such as facial recognition (FRT) and fingerprints into everyday use and has transformed interactions involving proof of identity. Personal devices—perhaps like the one on which you're reading this—have put that technology in the hands of the average service user and many of us are already granting access to our biometric data beyond banking, using it for online purchases and even workplace purposes. Although it may involve surrendering a degree of privacy, the adoption of this technology seems to be driven by the demand for administrative convenience. With the checking of identity, background and the absence of restrictions for leaving or entering being a source of 'inconvenience' for the traveller, the same interactions and multi-factor assurances needed in financial and legal transactions arise in the context of international air travel.

If we can reduce the inconvenience of international air travel by sharing our biometrics with the airport operators and carriers, that would seem to be an attractive option. Many of the administrative barriers to international air travel are imposed by the countries we are leaving and entering; sharing it with the airlines means sharing it with the state(s). As biometric data is highly sensitive, processing it (collecting, sharing, storing etc.) is closely regulated— where the processor

[4] As we discuss in Chap. 15.

is the state as opposed to, say, a bank or internet service provider, there are further important considerations.[5] Would these sensitivities stop us sharing our biometric data if we could reduce the inconvenience of international air travel? According to a recent passenger survey of air travellers in the United States,[6] the answer is a resounding 'no'. The survey showed that a substantial majority of U.S. travellers approved of the Transportation Security Administration's (TSA) use of biometrics at airports,[7] reporting 78% support among people polled by the U.S. Travel Association and market research company Ipsos. That might indicate that we're moving towards a convenience vs something trade-off, and a significant number of passengers (66%) supported the use of biometrics at airports if it was to make airports safer. And when we look at the reason that gained the most support (from 90% of passengers) for allowing use of the biometric data, it was the prospect of it getting them through security layers more quickly. At the same time, fewer than five per cent of respondents were concerned about their privacy while travelling. In other (admittedly less-than-scientific) words, *convenience* ranked highest and privacy the lowest.

Statistical rigour aside, the survey findings are also interesting for a number of other reasons, the first being the public attitude to AI-enabled technology when used by the state. We know that this is highly context-specific (Bayerl et al., 2023). In an extensive survey of over 10,000 UK citizens,[8] the majority (61.3%) supported the use of AI by the police but there was significant differentiation in use cases, with a high level of support where it is used to prevent terrorism and child sexual abuse, but the lowest level where it is automating administrative police functions such as call handling. The air travel survey results may be attributable to the context and not reflective of attitudes to any wider, non-security-related uses of our biometrics by the state. The peculiar risks and potentially catastrophic consequences addressed by air travel arrangements mean that neither they, nor the attitudes of travellers, are necessarily typical of the population generally. International air travellers have had to accept—or at least become inured to—intrusive searches of themselves, their luggage and belongings if they want to fly abroad. The attitudes may still be a question of 'security' as the payoff for sharing up biometrics. But once you've complied with all the state

[5] Chapter 10.

[6] https://www.biometricupdate.com/202409/airport-biometrics-one-thing-most-americans-agree-on.

[7] https://www.biometricupdate.com/202408/tsa-biometrics-program-is-evolving-faster-than-critics-perceptions.

[8] https://centric-research.co.uk/projects/aipas.

bureaucracy, international air travel presents you with a further inconvenience: waiting.

From a retail perspective, an airport is a shopping mall with a runway. Space control measures mean people are very unlikely to leave and have a lot of time to shop, eat, mingle and browse within a rigidly defined and confined area. And all of that can be made much more convenient if the traveller allows their biometrics—particularly their face—to be used by commercial entities operating within the airport. By giving airport operators advance permission to use their facial image, travellers can breeze through check-in and baggage drops, pay for duty-free goods, enter lounges and dine in restaurants using nothing more than a glance at a camera. The right personalised travel app will navigate around the airport using its facial recognition technology to steer them to specific outlets and ultimately to their boarding gate, even sending them nudge warnings that it's time to drink up and head for the departure gate before it shuts. There is little doubt that the 'convenience' being offered here is not solely that of the traveller. The ability for operators to maximise commercial opportunity, minimise cost and reduce loss is extremely attractive and several international airports offer this level of 'frictionless' engagement.[9] Whether people attending concerts and sports games would have the same attitude towards surrendering their biometric data has yet to be worked through, but this form of technological emancipation will revolutionise the experience of international air travel. With pre-travel identity verification, digital IDs, and digital travel credentials (DTCs) allowing optimisation of administrative processes and enabling real-time risk assessment, this will also revolutionise border enforcement.[10]

But what about the 'security' side of the equation? The evolution of biometric data processing at airports raises a compelling question: for what purposes should the state be able to use the data from such a highly advanced form of remote biometric surveillance? From the strictly security point of view, live FRT systems in an airport can get us on our flight quicker, more cheaply and without unnecessary contact with each other, and it makes no sense not to use the same system to keep the plane in the air and to prevent the flight being used by those with malign intent—in fact, not doing so would be legally and ethically dubious. Would the same security interests apply in less rigid and controlled settings like shopping centres and concerts? Just because the air passenger is prepared to

[9] https://www.biometricupdate.com/202502/beyond-borders-biometric-driven-efficiency-in-travel.

[10] https://www.biometricupdate.com/202502/beyond-borders-biometric-driven-efficiency-in-travel.

allow their facial image to be used at check-ins and check-outs, will that mean that someone booking a domestic train journey will be content for their image to be used by state agencies for any and all non-security purposes like tax collection or reminders about vaccinations or voter registration?

Even in airports it is unclear what will happen where there is a wider public interest in identifying the passenger and their movements. The U.S. survey showed that air passengers supported the use of their biometrics for combatting terrorism and human trafficking, but those extreme scenarios are generally easy ones to test with the public. Will travellers be equally content for their journeys to be minutely tracked by the police who are looking, not for a terrorist or trafficker, but a competent adult who wants to find some solitude but has been reported as 'missing' by a concerned relative[11]? Personal convenience solutions generally rely on express prior consent (admittedly skipping the small print) before biometric data can be used, with the availability of opt outs, conditions and less intrusive alternatives being a requirement of many legal frameworks.[12] When the citizen agrees to their biometrics being processed for travel (by a combination of commercial operators and law enforcement agencies like border and customs control), how far can the state infer consent for it to be used in other public interest purposes? If the police from another country—not the traveller's destination—ask the UK for help in locating a 'person of interest' for political, occupational or economic purposes, can they use the biometric data that the person voluntarily—and conditionally—shared with a travel agent to do so? In both examples it would be easy for the police to contend that there was a proper policing purpose like a risk to life or diplomatic requirement in locating the person, but it would be stretching things to say they were a serious 'security' issue. Any justification for processing biometric data that had been shared for traveller convenience would come from a wider public interest—where will those lines be drawn and by whom? Regulators in countries like Norway are now pushing for clearer lines in the widening use of biometric identification with outright bans on some applications.[13]

The convenience/public interest boundaries also need technological lines to be drawn. Airport biometric systems have now "matured into a hybrid system that pairs airline-installed cameras at boarding gates with government-issued handheld

[11] For the growing challenges of going 'off grid' see Chap. 15.

[12] For example the European Union's General Data Protection Regulation, https://gdpr-info.eu/.

[13] https://www.biometricupdate.com/202510/norways-data-privacy-watchdog-seeks-ban-on-remote-biometric-identification.

devices carried by federal officers".[14] Where airlines or airports lack their own biometric infrastructure, Customs and Border Protection (CBP) officers are using mobile devices to fill the gaps and have reportedly become a 'permanent feature of departure control' in the United States. This hybridisation is blurring the lines between 'convenience biometrics' and other traditional forms of surveillance being conducted at exit gates, creating concern among passengers, with *The New York Times*[15] reporting how one at Boston's Logan International Airport described departing international travellers being "ambushed" by CBP officers with what appeared to be cell phones.

Convenience for suppliers is clearly a driving force behind the technology being offered, sometimes overtly and sometimes packaged as a mutuality of convenience—just try to avoid moving to an app rather than using established website visits and you'll feel its gravitational pull. The consumer convenience offered by biometric technology can be mutually attractive in settings beyond the rarified setting of airports, but the level of exposure to state surveillance based on public interest that this would bring for the citizen would be exponentially greater. There are some obvious conveniences in allowing my facial image to be used when buying a ticket for a fast-selling sporting event. In addition to the ease, speed and certainty of purchase, using a biometric identifier will allow me to pass straight through the turnstile, to locate my seat and ensure it isn't already occupied by someone who bought a counterfeit ticket outside. But should that image be available to the police for live or retrospective facial matching[16] if there's a serious incident at the event? What about if there's an outbreak of disease traced back to that gathering? For how many years after the event must my biometrically verified image remain accessible to the state? The police are already operating live FRT at large public celebrations and other unticketed, crowded space events like the Notting Hill Carnival (Sampson, 2025), so why shouldn't they be allowed to access reliable biometric data already captured by the event organisers in the name of convenience if it serves a clear public interest? Employers are increasingly recognising the benefits of offering biometric access to offices and workplaces on a voluntary basis, but there are now moves towards making the use of facial or iris matching compulsory (Martin, 2025); it will be only a very short matter of time before the police want access to that biometric data.

[14] https://www.biometricupdate.com/202509/cbp-biometric-exit-blurs-lines-between-convenience-and-surveillance.

[15] https://www.nytimes.com/2025/09/26/travel/airports-biometric-exit-program.html.

[16] For an explanation of what this is and how it works, see Chapter 6.

Developments in international travel show how we are moving the legal and societal challenge from a privacy: security equation towards a consideration—or even conflation—of convenience and public interest. The changes in our behaviour, expectations and tolerance of intrusion, more generally, are raising many surveillance questions, shifting the situation but not yet the discussion. Commercial organisations are openly talking about 'the consumerisation of biometrics', offering eGates and liveness detection pods beyond the travel environment, with their sights on sectors like healthcare and real estate.[17]

Whether a 'private-or-safe' dichotomy was ever an accurate representation of the competing issues, remote biometric and other AI-enabled surveillance capabilities are becoming less about a bipolar face-off and more about perpetual balancing. And the dynamic elements being balanced are increasingly personal convenience and public interest.

Those given more to cynicism in this field might say that, whatever we call it, this is a suspiciously circular argument and one that Silicon Valley might recognise. By first creating inconvenience in the name of security, the state can then introduce highly intrusive technological solutions which would otherwise attract more scrutiny and resistance. I am content to leave those arguments to others. Suffice it to say that the emerging evidence corroborates a move away from a straight-line trade-off between privacy and security and a future where balancing personal convenience with public interest will be guiding any line drawing for AI-enabled surveillance.

References

Bayerl, P. S., Obst, M., & Akhgar, B. (2023). *International citizen consultation on AI accountability in policing* (AP4AI Report 3).

Martin, B. (2025, October 10). JP Morgan makes biometric data compulsory at new HQ. *The Times*.

Sampson, F. (2025, August 4). From Notting Hill to Denmark Hill—UK Police facial recognition rejoins the carnival. *Biometric Update*.

[17] https://www.biometricupdate.com/202510/clear-charts-path-from-consumerization-of-biometrics-to-cross-sector-platform.

8 Perpetrator Focus—A Culture Shift in Surveillance

Focus on the perpetrator. That was the message at the start of freshers' week for universities in October 2024 from the UK's newly appointed justice minister.[1] Alex Davies-Jones was talking about tackling the spiking of drinks in bars and clubs and urged us to make a 'culture shift' so that the focus was on preventing and deterring the perpetrators' harmful and unwanted behaviour, rather than expecting everyone else to adapt theirs. Intuitively this sounds right and surveillance practice in licenced premises has been very much in that direction for some time—now the technology is following.

If we're trying to protect people—whether staff, students or customers—stopping the assault, the abuse or the harassment from happening in the first place is surely the best surveillance outcome. The best measure of police effectiveness is not the amount of police activity but the absence of crime.[2] By the same token, the practical success of AI-enabled biometric surveillance in the retail sector has come from the dramatic drop in incidents of violence and abuse of staff.[3]

Heading into the future of AI-enabled surveillance, if we were to ask shop workers about their experiences, about vulnerability and what they would like their employers to prioritise when considering new technological solutions, they might take the same view. The delivery drivers who are spat upon so often

[1] https://www.theguardian.com/uk-news/2024/sep/22/rewrite-victim-blaming-drink-spiking-campaign-uk-civil-servants-told.

[2] https://www.biometricupdate.com/202507/how-to-measure-success-in-biometric-security.

[3] Chapter 11.

F. Sampson, *How AI-Driven Surveillance is Changing our World*, Security Informatics and Law Enforcement, https://doi.org/10.1007/978-3-032-24005-7_8

that they are being issued with DNA kits[4] might agree with the former minister that the biometric technology of the future should be *protecting* them from the perpetrator's harmful behaviour rather than responding in tacit acceptance of it happening. Surveillance measures that catch culprits, collate evidence and support prosecutions are an essential part of integrated security systems and are a key source of support for the police. But one of the most potent features of AI-enabled biometric surveillance technology in the setting of retail and hospitality is its ability to prevent crime in a way that has proved impossible—or at least impossible to prove—in the last 50 years of CCTV.[5] And, while policymaking is for others, there is good practical news for current ministers thinking about changing surveillance culture to being more perpetrator focused: we've been here before and it worked.

Several decades ago, the UK introduced a voluntary security scheme where known perpetrators of behaviour such as drink-spiking were identified, banned from local licenced premises and occasionally prosecuted. The successful Pub Watch scheme was aimed primarily at prevention and was strongly supported by the police. It relied on several features. First, licensees needed to identify the behaviour that was blighting their sector. The lists were fairly consistent across the sector: sexual harassment of staff and other customers, the selling of controlled drugs, stealing from coats and handbags, avoiding payment for their drinks/meals and physically or verbally attacking people—including door security staff.

Once they had identified the behaviour, the licensees and their staff had to identify the individuals who they knew or believed were responsible for it. Often, the small number of perpetrators were known to people who worked in the premises and could be identified by them; in many towns they were the same people. The landlords put the names and photographs of these 'frequent flyers' on a local list and shared them with fellow members of the scheme.

If an individual who was on their list tried to enter their bar, the licensee could intercept them at the door and either warn them to behave appropriately or deny them entrance altogether. Where they barred anyone from entering, the licensee would then send a photograph (in the early days, by fax machine) to other bars and premises in the area so that they would anticipate the person showing up at their door and could decide how to deal with them if they did.

---

[4] https://www.grocerygazette.co.uk/2024/08/19/tesco-drivers-dna-tests/.

[5] Chapter 5.

This simple perpetrator-focused scheme for pubs had some obvious flaws, the first being its dependence on the ability to identify the right person. Notwithstanding their experience and diligence, bar staff (particularly at busy times) have the same human frailties as the rest of us. When it comes to visual identification—even by police officers—there are many criteria that can impair someone's ability to reliably pick out a person accurately.[6] Distraction, obstruction, noise, fatigue, eyesight. Trial lawyers are very aware of these limitations and can often focus a lot of their questioning of witnesses in this territory, so much so that in England and Wales there is a schedule of them. To help remember what they are, the police and prosecutors often arrange the factors into the ADVOKATE acronym which looks like this:

ADVOKATE factors

- **A**mount of time: for how long did the witness observe the suspect?
- **D**istance: how far was the witness from the suspect?
- **V**isibility: what were the lighting conditions like at the time of observation?
- **O**bstructions: was their view blocked by things or people?
- **K**nown or seen before: had the witness previously seen the suspect?
- **A**ny reason to remember: is there a specific reason for the witness to have a clear memory of the event or the suspect?
- **T**ime elapsed: how much time has passed since the event occurred?
- **E**rrors: were there any discrepancies between the initial description and the person's appearance?

In criminal trials where visual identification is in issue, these factors are so important that judges must warn juries about the dangers of convicting someone on visual identification evidence.[7]

The relevance of all this to the pub watch scenario can be seen very clearly. Even where the door or bar staff know the individual (as opposed to having only seen their image on a sheet of faxed paper or sometimes a verbal description passed over the phone), the environment of the night-time economy, with all its ambient lighting, crowded spaces, to-ing and fro-ing of people, is less than ideal for getting a clear, unobstructed view. If any such ideal conditions existed, there is also a general human unreliability underlying the *Turnbull* case and the fact that we are far from infallible when recalling detail (Otten et al., 2023).

---

[6] Note these external, environmental factors also apply to 'super recognisers', people with an innate ability to recall faces.

[7] Known as the Turnbull guidelines, from the case *R v Turnbull* [1977] QB 224].

Nevertheless, pub watch was a successful early surveillance scheme and the increased use of CCTV improved it further. With cameras offering better coverage of entrances and parts of their premises, plus the ability to share images of the individuals, licensees could prevent people entering their private licenced premises in a town or neighbourhood for a given period, sparing police resources while avoiding harm and expense.

Since the original pub watch scheme, the technology behind AI-enabled surveillance has been transformed beyond recognition—all the more surprising then that, in the retail sector, the methodology is the same.

The numbers involved in retail crime are astounding[8] but the challenge is simple—how to stop £millions walking out of the door and nasty people walking in. When weighing up the surveillance issues in retail crime it's worth remembering that most of our nation's shops are private premises. Like pubs and clubs, we enter shops by implied invitation. Like the pub watch scheme, if the owner knew we planned to steal from them, attack staff or other invitees, they would withdraw that invitation—and the law in England and Wales would let them. Unlike licenced premises, the volume of visitors to supermarkets is almost impossible to monitor continuously. With high footfall, multiple entrances and exits and fewer opportunities for staff to identify known perpetrators, even those supermarkets and large stores with modern CCTV systems have struggled to tell the few people who should be excluded from genuine customers. The most effective way to stop people doing unwanted things in your shop is to withdraw the invitation of the perpetrator, the problem has been knowing who to let in and whose presence deserves further scrutiny.

Having realised that their cameras aren't working, some retailers have turned to biometrics. With stores reporting reductions of up to 70%—not just at the start but as a sustained outcome—live facial recognition (LFR) technology is helping retailers take back control of their businesses. Why? Because LFR is perpetrator focused, empowering staff to deny entry to individuals who have previously caused them harm or loss; they can do it quickly and confidently and it's having extraordinary results. Unlike CCTV cameras, biometric surveillance can be individualised. Live facial recognition capability is perpetrator focused, allowing identification to be carried out with far greater speed, accuracy and scale, no longer relying on word of mouth and swapping photos, but using algorithms, auditably and accountably, with clear records, policies and data sharing protocols.

---

[8] https://www.biometricupdate.com/202508/why-retail-needs-biometrics-the-cameras-arent-working.

Elsewhere on the high street retailers are starting to use the same combination of focus and technology, deterring the would-be offender rather than expecting everyone to accommodate rapidly rising retail crime. In the era of AI-driven security, crimes against staff, customers and stock are being prevented, and it's offender-focused technology that's stopping them.

Another feature that retail surveillance shares with the British pub is the limited number of perpetrators. Retailers estimate that 70% of thefts from their stores are committed by around 10% of shoplifting offenders. Admittedly, 10% of all shoppers is a big figure, but the overall number of people responsible for UK-wide shop crime was put at 29,000.[9] When considered against the number of retail crimes and the loss to the economy,[10] this is a small and disproportionately impactful population.

Unaffected by the size and scale of the challenge, undistracted by environmental factors and unwavering for as long as they have a power supply, remote biometric surveillance systems will monitor all visitors to all premises, consistently and with near-perfect levels of recall. Because they 'know' who they're looking out for, they are able to discriminate and detect specific faces in large crowds, even where they only have a partial view or the person is wearing a head covering or mask. Where CCTV cameras catch everything in view and retain a copy for someone to sift and discard later, live FRT systems don't. If they're not looking for you, the facial matching systems don't 'see' you, and don't record you. Compared in this way, CCTV is bottom trawling; FRT is spear fishing. As they have no information about you, FRT systems cannot 'recognise' or identify you. And, with accuracy ratings higher than 99.99%, they are very unlikely to need the ADVOKATE acronym. In short, they are more accurate, less subjective and vastly more reliable than security staff assisted by conventional CCTV systems.

A key question when encountering these systems is how you can become known to the AI-enabled device. This is similar to the creation of the perpetrator lists compiled by pub watch members. Providers of facial recognition surveillance systems maintain a database of individuals who have been caught committing offences in members' stores or who are strongly suspected of having done so based on their earlier behaviour. Images of these individuals are fed into the algorithm which then scans faces of people entering the premises and sends a

---

[9] https://www.standard.co.uk/news/crime/criminals-shoplifting-crime-public-purse-billions-b1250596.html.

[10] £2.5bn in the UK in 2024/5—https://brc.org.uk/news-and-events/news/corporate-affairs/2025/ungated/retail-crime-spiralling-out-of-control/.

notification if there is a match above a certain threshold that is set by the user. The match is checked by a facial-matching expert and the shop staff are alerted. In this way the algorithm is doing the looking and matching, and the shop staff are intervening and deciding which action to take in respect of the individual, who invariably chooses to leave the store.

State-of-the-art surveillance has come a very long way since the introduction of CCTV, when images—even from police and local authority cameras—were so grey and grainy, they could have made equal claim to being sightings of the Loch Ness Monster as a shoplifter or assailant. Technologically, that dinosaur period is now part of British folklore too.

Retail premises like bars and shopping centres that have traditionally attracted certain types of offending and offenders can now access AI-enabled surveillance capabilities which literally focus on the perpetrator, while perpetrators quickly learn this and know which premises have biometric security systems that are looking for them.

Policing isn't shopkeeping and the challenge for the state is, of course, much more complex than denying some people access to commercial premises. It is too early to tell how far the success of biometric surveillance in retail venues will shift perpetrators to other, less protected targets, but for the police, displacing crime doesn't solve it. Solutions like LFR empower retailers and licensees to take back control of their commercial space and identify the small minority of people whose previous visits to their stores have resulted in loss, assault, intimidation and predatory behaviour. But most private shopping malls and bars are also 'public places' for the purposes of policing,[11] so too are train stations, bus depots, airports, sports and entertainment venues, all of which need some policing input. The different goals and intended outcomes for the retail sector and the police should be borne in mind at the very outset of any proposed joint biometric surveillance ventures.

The use of AI-enabled surveillance in the retail setting is expanding rapidly and is certain to spread further,[12] but even in that very focused setting, reflection and restraint are needed.

The deployment of LFR in order to protect food stores has been expressly endorsed in New Zealand, where the Privacy Commissioner noted[13] the need for

[11] See the Public Order Act 1986, s.16 (b) "any place to which at the material time the public or any section of the public has access, on payment or otherwise, as of right or by virtue of express or implied permission".

[12] https://cisp.cachefly.net/assets/articles/attachments/50669_to_boom.pdf.

[13] https://www.biometricupdate.com/202506/new-zealands-privacy-watchdog-okays-foodstuffs-facial-recognition-trial.

retailers to approach the technology with caution and care. If they are to use this technology lawfully and responsibly, UK supermarkets and other retailers must factor in some major considerations before they even think about adopting remote biometric systems. Facial recognition isn't a plug-and-play solution and there is significant risk in cross-breeding technologies.[14] Choosing a competent technology partner is essential, as the potential pitfalls can be catastrophic.[15]

Ministers in several countries have recognised the power of AI-enabled surveillance technology for the police, with those in the UK tending to see it bringing swifter and surer justice to offenders in the aftermath of the 2025 public disorder centring on immigration. That same technology can enhance policing capability in sectors such as retail and licencing by preventing high volume offending against a backdrop of prosecution backlogs and a burgeoning prison population. With perpetrator-focused solutions, advisers must be careful to balance competing rights, attentive to data protection and privacy requirements, the chilling effect of surveillance generally and the ever-present risk of function creep. Having clear legal and procedural safeguards, together with comprehensive policies, audit processes, training plans and complaints protocols, will help mitigate the risks and reinforce the need to demonstrate the necessary public interest in use of new technology.

Following the formation of a new government in 2024, the chair of the UK's Home Affairs Select Committee avowed that they would 'not tolerate the criminality blighting our communities'. Zero tolerance is an easily stated strategy and has often involved putting up notices declaiming certain types of unacceptable behaviour. This kind of signalling has been fashionable in shops and licenced premises for a while, but the crime numbers suggest any 'intolerance' has been extremely elastic. Properly designed and managed surveillance systems mean retailers—and ministers—can now back up their words with their technology and prevent the behaviour that is undeniably blighting their sector. While not all are using biometric options, most retailers have exhausted the other methods and their overall crime rate still features in the alarming statistics.

The next time you're in a bar or shop, look around for notices claiming the organisation does not tolerate behaviour like assaults or abuse. If allowing something unwanted to happen without available interference is used as the definition, some places have been extremely tolerant. AI-enabled surveillance technology

[14] Chapter 5.

[15] In the UK the data regulator can fine companies £17.5 million or 4% of the total annual worldwide turnover in the preceding financial year, whichever is higher—https://ico.org.uk/for-organisations/law-enforcement/guide-to-le-processing/penalties/.

means organisations can now intervene and really focus on the behaviour they say they're trying to prevent. That's a culture shift.

Reference

Otten, M., Seth, A. K., & Pinto, Y. (2023). Seeing Ↄ, remembering C: Illusions in short term memory. *PLoS ONE, 5*, e0283257. https://doi.org/10.1371/journal.pone.0283257

Done Nothing Wrong? The Worst Surveillance Advice Ever Given

9

Asking 'what's wrong with any of this?', this chapter provides a forensic deconstruction of the premise that, if you have done nothing wrong, you have 'nothing to worry about' from what has been discussed in other chapters. With examples of why this is a fundamentally flawed approach in the world of AI-driven surveillance, the chapter debunks the long-held view of surveillance, noting how the fallacious argument that only the guilty need fear the use of intrusive technology is still frequently deployed by law enforcement and governments as an argument for intrusive surveillance.

> If you have done nothing wrong, you have nothing to worry about.

I still hear this advice a lot. When I hear people offering it in a surveillance context, I smile. When those people are government ministers, I wince. And when they are police chiefs I begin to despair.

It's often deployed more as a defence than an explanation, and usually part of a wider 'security vs privacy trade-off' (see Solove, 2021) but it has no place in the surveillance present, let alone the AI-enabled future.

The realities of surveillance experience ought to illustrate why this is so, and yet the 'nothing to worry about' fallacy has been deployed as a basis for introducing, expanding or explaining intrusive public space surveillance for decades and it needs to be called out.

When we look at the technological, legal and societal issues closely, any logic behind this argument collapses, revealing that it is neither a valid defence nor

F. Sampson, *How AI-Driven Surveillance is Changing our World*, Security Informatics and Law Enforcement, https://doi.org/10.1007/978-3-032-24005-7_9

mitigation to public surveillance.[1] It is such a stubbornly resistant position in debates about the police use of AI-driven surveillance capabilities, I felt compelled to publish a deconstruction of this fundamentally flawed premise from a wider legal perspective in advance of giving evidence to parliament.[2]

Taking some cues from a brilliant de-constructor of illusory pseudo-logical argument,[3] the flaws in the 'nothing to worry about' advice go like this:

First, it is based on a presumption of guilt. It doesn't matter whether you have done anything wrong; in order for it to be true, the advice requires you, *the citizen,* to *prove* you have 'done nothing wrong'. In most democratic jurisdictions this is a reversal of constitutional legal safeguards and runs counter to one of the most important in any decent criminal justice system: the presumption of innocence. For those of us lucky enough to live in countries where we begin the criminal justice journey as innocent until the state has proven otherwise by meeting the relevant evidential standard, this presumption is critically important.[4] Running the argument that only the citizen who is able to show a clean pair of heels faces no risk from surveillance effectively allows the State to pass the burden of proof back to you. Not only does this undermine any constitutional presumption of innocence, it also relays a clear constitutional message—that privacy, freedom of movement, speech, assembly and other cherished democratic entitlements are only privileges, enjoyed (and therefore withdrawn) entirely at the pleasure of the state. This inference goes against established international and constitutionally protected rights. Reversing the state v citizen relationship in this way is directly relevant to the area of police surveillance, as illustrated by several legal challenges to the police maintaining details and/or images of unconvicted persons on searchable databases which we will come to below.

The further flaw in the 'show us you've done nothing wrong' requirement by the police is that it puts the citizen in the notoriously difficult position of

[1] https://www.amnesty.org/en/latest/campaigns/2015/04/7-reasons-why-ive-got-nothing-to-hide-is-the-wrong-response-to-mass-surveillance/.

[2] Evidence of Professor Fraser Sampson to the Joint Parliamentary Committee on Human Rights (22 February 2023). https://committees.parliament.uk/event/17397/formal-meeting-oral-evidence-session/.

[3] Carl Sagan's Baloney Detection Kit: https://www.openculture.com/2016/04/carl-sagan-presents-his-baloney-detection-kit-8-tools-for-skeptical-thinking.html#google_vignette.

[4] For those who don't live in such countries, the risks of surveillance will already be obvious.

having to prove a negative. How would you prove, for example, that you are not dishonest[5]?

The second major flaw in the advice comes from the source of information we and the state are going to rely on when deciding if you've done nothing wrong. This technological issue can be split into two parts, the first being database infallibility. (computer says "wrong"). Increasing reliance on databases and data retrieval has resulted in deference to database entries—or, increasingly, algorithms. This fact of techno-driven life means that, if the computer says the citizen has done something in the past, then they have 'done something'. This is not just a policing issue and applies to any administrative record that the state (in the form of a hospital, passport office or school) relies on for its 'truths'. Efficient systems rely on computer records and if that computer is operated by a law enforcement agency, its verdict is treated a bit like a breathalyser showing red: computer says you've been drinking—prove you haven't. If the computer is running a police algorithm looking for a *specific individual* who is a known criminal and it 'recognises' you as being that person, there's an irresistible inference that you're a criminal. Already faced with having to prove a negative above, the your burden becomes exponentially harder once you have to disprove the entries on the state's 'trusted' database in order to do it. This risk is at the heart of data protection rules such as the need for human intervention in certain types of automated decision-making (and the accuracy of records.). The second part of this problem is database *fallibility* (computer is wrong). Sometimes it is *the computer* that is 'doing something wrong'. In the AI-enabled future of surveillance there is a significant risk that we will increasingly be judged by algorithms and find we have to meet a standard far higher than the algorithm was itself. The accuracy of police surveillance algorithms has increased significantly in recent years but there remains a risk of confusing precision with reliability. The citizen should not be required to disprove an entry on the state's trusted watchlist database, still less have the burden of *discrediting* the system itself.

A third feature of the 'nothing to hide, nothing to fear' approach is that it presents two fatal outcomes and invites us to pick one. Accept or admit. Like scamming or cheap political activism, the argument imposes a false dichotomy and then assumes the answer, excluding any shades or nuance in between.[6] It assumes

[5] NB the absence of criminal convictions for offences involving dishonesty isn't much of an argument.

[6] Like the armchair advocate insisting on a 'yes or no' answer when the question obviously cannot be answered that way "Have you stopped stealing – Yes or No?".

that all state intrusion is benign because the only time you need worry about it is when you've done something wrong.

Given the considerable difficulties in proving the negative and disproving the police computer, it gets worse. Once the police computer has determined that you have 'done something wrong' and you are prosecuted (using the wholly flawed algorithmic records), your lawyer may advise that your best tactical option at trial is to plead guilty and at least get some discount in your sentence or plea deal. By following the proper and sound legal advice, you admit culpability and get a criminal conviction, thereby vindicating the 'done something wrong' argument. Adding insult to ignominy, this fate is precisely what happened in the prosecution of the UK postal workers[7] a cautionary tale which on its own ought to put an end to the fallacy once and for all but it hasn't yet done so.

A fourth flaw in the advice (if the others weren't compelling enough) is its appeal to innocence. Opposing the advice, or challenging the argument, is culturally seen by law enforcement as self-incriminating. Arguing against its (flawed) logic is itself a sign of having something to hide. In the US Supreme Court case that gave American citizens the right to silence,[8] Chief Justice Warren said suspects needed a right to have a lawyer present because of the police assumption that not answering their questions meant you have something to hide. In other words, in the eyes of the police resisting the argument supplies the corroborative evidence of 'wrongdoing', and the 'nothing to hide nothing to fear' argument is triggered by the very reliance on the Fifth and Sixth Amendment rights given to them by the Constitution.

In addition to the technological, legal and logical failings of the advice, the 'nothing to hide' fallacy suffers from chronic generality. What exactly do the police mean by 'wrong' when they use this expression? Does speeding count? How about non-criminal behaviour like an officer suspected of 'pulling a sickie' from their employer? What about the people who technically may have breached their countries' COVID-19 lockdown rules by meeting up or exercising or worshipping[9]? Have they 'done something wrong', something that means they are fair game for any surveillance? Is the test a legal or ethical one? Humanity errs and, without being specific the 'done *nothing wrong*' assurance leaves no one to be consoled by absolute innocence. And if we're using criminal convictions as

---

[7] S Ravikumar, 'What is Britain's Post Office scandal?' (*Reuters* 11 January, 2024). https://www.reuters.com/world/uk/what-is-britains-post-office-scandal-2024-01-09/.

[8] Chapter 12.

[9] Chapter 19.

the yardstick, has someone found guilty of a serious offence by who then has their conviction quashed because of a procedural error, 'done something wrong'?

In addition, rules and standards (legal and ethical) change after technology has been introduced. Things that were not 'wrong' when surveillance systems were approved and deployed may subsequently become 'wrong' (aka '*function creep*' which has been shown to increase with synchronisation of state functions and databases). These developments are rarely accompanied by new rights and, if the citizen surrendered all their old ones, there are unlikely to be any more coming their way. Looking in the other direction, criminal laws are amended in line with societal attitudes and things that were once seen as being 'wrong' and for which many people will have been convicted, are no longer treated as such. Possibly the best examples of this phenomenon are sexual offences. In 1990 a man in England and Wales could not be convicted of raping his wife even he admitted to forced sexual intercourse without any form of consent,[10] while two adult men engaged in consensual sexual activity in the presence of another adult committed an offence punishable by two years' imprisonment.[11] Which of these might have fit the 'done something wrong' criterion?

As an idea, the 'done nothing wrong' fallacy is easily defeated, but as a policy position it would present an almost insurmountable hurdle for the citizen. Using the argument rhetorically as a debating tool in parliament[12] is one thing, but the literature on intrusive surveillance by the state (See e.g. Anderson et al., 2018; Joern, 2009) and other state activity such as human genetics[13] completely and consistently debunks this approach, but the argument that only the guilty need worry is still deployed in formal political debate.[14]

How it has survived so long in policing is difficult to discern but there are a couple of clues that are specific to a law enforcement context.

The first is the premise that the world is divisible into two unequal parts: the 'law abiding majority' and everyone else. In the UK this premise can be heard beneath many policy discussions and is deployed by home secretaries and

[10] See *R v R* [1991] UKHL 12.

[11] Sexual Offences Act 1956, s.13 and Schedule 2.

[12] https://hansard.parliament.uk/Commons/2020-06-24/debates/062FA715-C506-4F43-9F2B-0DBDF51331C3/WestferryPrintworksDevelopment.

[13] Nothing to hide, nothing to fear?, Human Genetics Commission report, Nothing to Hide, Nothing to Fear? November 2009 p 22. http://www.hgc.gov.uk/UploadDocs/DocPub/Document/Nothing%20to%20hide,%20nothing%20to%20fear%20-%20online%20version.pdf.

[14] https://hansard.parliament.uk/commons/2023-06-27/debates/1557FA76-F69E-4E14-B459-10BD7BBB0917/RepurposingRussianAssetsToRebuildUkraine. https://www.thejournal.ie/surveillance-bill-goebbels-2426368-Nov2015/#.

police chiefs. While one would hope that the numerical majority of our fellow citizens will acknowledge the Rule of Law and doesn't represent a threat to each other, their property or the King's peace, the 'law abiding' bit suffers from the same imprecision above. A regime that tolerated no deviation from any of its many laws and treated any individuals who breached them as outlaws would be unrecognisable to most of us.

So, who do we believe makes up this 'law abiding majority' in whose name we are acting, and who is excluded from any consideration?

The second law enforcement-specific thing keeping this holed advice afloat might be the privacy paradox for policing.[15]

The citizen is deploying increasingly intrusive surveillance technology like smart doorbells and IoT devices in every aspect of their, happily surrendering sensitive personal data to private companies, the police and each other but also sharing it with the police themselves. What could such a population of consensual 'digital naturists' possibly object to if their data was used by the police to keep them and their loved one safe? The law allows the police to take and keep photos and DNA samples and fingerprints, to track millions of vehicles per minute on the UK's roads[16] and require the removal of face coverings in public, all in the name of protecting the citizen from serious harms, so why would they object?

As surveillance technology transforms our world, our understanding of how it needs to be regulated must also undergo transformational change. Police AI will bring them extraordinary powers and new capability, and any effective regulatory framework will need to enable the accountable use of both. A recent UK survey of over 10,000 people and their attitudes to the police use of AI found that a major concern among respondents was that the police would blame their algorithm for what were their own failings in the future (Akhgar et al., 2025). This is highly significant in the context of surveillance intrusion and very interesting when seen against the government's 'defence' to unlawfully retaining millions of images of people who had been arrested but never subsequently proceeded with.[17] In the future, the 'computer was wrong' element of the 'done nothing wrong' fallacy will become more prominent. In addition, 75% believed that there should be mandatory accountability processes in place before the police use AI tools and only 20% believed that accountability mechanisms were in place.

---

[15] See Chap. 18.

[16] With an officially 'acceptable' error rate of approximately 4 million reads *per day*—see Chap. 13.

[17] The Home Office blamed the lack of a bulk delete capability in the database which they had built an argument that found little support from the court in Catt v. the United Kingdom (Application no. 43514/15).

We consider police accountability for their use of AI in another chapter but for our purposes here we should note the increasing practical importance of knowing what to worry about from state surveillance, whether we've 'done anything wrong' or not.

If you hear anyone suggesting that 'so long as you've done nothing wrong, you have nothing to worry about' from AI-enabled surveillance, treat it, not as wisdom but a cry for help because the person saying it is obviously and hopelessly out of their depth.

References

Akhgar et al. (2025). *Survey*.

Anderson, J., Skare, E., & Dorroll, C. (2018). Nothing to hide, nothing to fear? Tools and suggestions for digital data protection. *The Qualitative Report, 23*(5), 1223–1236.

Joern, L. (2009). Nothing to hide, nothing to fear? Tackling violence on the terraces. *Sport in Society: Cultures, Commerce, Media and Politics, 12*(10), 1269–1283.

Solove, D. (2021). The myth of the privacy paradox. *George Washington Law Review, 89*, 1.

Who Says the Police Should Be Using AI?

10

Despite constitutional differences, police chiefs in many countries are united by two practical questions: 'What does accountability for police use of AI-enabled surveillance look like?' and 'How do we achieve it?' The chapter answers both.

Debates about the potential use and abuse of Artificial Intelligence (AI) continue to make headlines as we come to terms with its reach. How to adopt and adapt to AI responsibly and accountably is a strategic challenge for every sector and policing is no exception.

Artificial Intelligence is bringing extraordinary surveillance capabilities to policing. Other chapters in this book illustrate where the technology has already taken us, how it is shaping our lives and vice versa. It will also bring accountability.

Accountability is about people with power using it properly, and, where those people are the police, their use of power can have profound consequences. If their power is amplified by technology, accountability extends to its use. Advanced technology comes with advanced accountability, and the state will need to answer the question 'who watches the watchdogs' AI?'.

While it includes an obligation to explain,[1] police accountability is not so much about having *the* answer and more about having *to* answer. The ways in which the police have to answer for their action (and *in*action), and the people or organisations to whom the police are answerable vary significantly from one country to another—it even differs across the local policing jurisdictions in the UK. However, the *practical* ways in which they might use AI-enabled technology to prevent and investigate crime are very similar across the world. Despite any

[1] This is particularly the case in relation to 'explainable AI'—see Chap. 1.

F. Sampson, *How AI-Driven Surveillance is Changing our World*, Security Informatics and Law Enforcement, https://doi.org/10.1007/978-3-032-24005-7_10

constitutional and legal differences, with AI-enabled solutions becoming more affordable and reliable, police chiefs in many countries are united by two practical questions: 'What does AI accountability look like?' and 'How do we achieve it?'

These questions were at the heart of an ambitious research project conducted by researchers at the Centre for Excellence in Terrorism, Resilience, Intelligence and Organised Crime Research (CENTRIC) (Akhgar et al., 2022)—and the full answers are set out in the subsequent AIPAS framework published in 2025 (AIPAS[2]).

Accountability is closely aligned with lawfulness, but there's more to it than keeping within the law. Police organisations are emanations of the state and exercise exceptional powers. Insofar as legality is concerned, a body of law is beginning to develop around the use of intrusive police surveillance technology, the storage of data and so on,[3] but there are situations where the technology has fundamentally altered the relevant considerations. For example, AI-driven retrospective facial matching capability being rolled out across UK policing bodies fundamentally alters the ability to search, compare and combine data points from which individuals can be recognised in perpetuity. This makes the level of potential intrusiveness wholly and materially different from any technology that might have been considered by the courts previously. The courts in England and Wales have held that the novelty of technology 'does not mean that it is outside the scope of existing regulation',[4] or that it is always necessary to create a bespoke legal framework for it.[5] However, while new technologies will be able to integrate into the co-existing common law rules that govern policing and surveillance more broadly, the extent of novelty in AI-powered technology may mean there is no precedent to apply.

With AI-enabled capability, we cannot always seek the comfort of doing what has worked before because we have not been here before.

Operating under the Rule of Law, the police are answerable to the courts for their use of AI tools, but there are other pathways that need to be in place in order to map their wider accountability.

In countries like the UK that have a policing model based on democratic government, citizen support for the use of AI-enabled capabilities will be of central importance. Even in countries with a more detached, authoritarian policing model,

---

[2] https://aipas.co.uk/.

[3] *R (on the Application of Catt) v Commissioner of Police of the Metropolis and Another* UKSC/2013/0114.

[4] *R (on the application of Bridges) v Chief Constable of South Wales* (2020 EWCA Civ 1058).

[5] *Bridges* at 84.

the police will need the support of the citizen in using AI-enabled surveillance because so much of the material on which the state relies will increasingly come from sources other than police-operated cameras.[6] Central to the state gaining and retaining the support of the citizen will be accountability.

The starting point for the study was to disentangle general principles from legal requirements and identify the practical differences between broad accountability and narrow legality. Accountability is closely aligned with lawfulness to the extent that other features are Lawfulness *Plus*. Policing bodies are emanations of the state and exercise exceptional powers. They usually enjoy extraordinary privileges when compared with other organisations and, because they interfere with the fundamental rights and freedoms of the citizen, must be able to identify their legal authority to act. As such, police bodies operating under the rule of law are answerable to the courts. But that is only part of police accountability and there are other pathways that need to be in place in order to map accountability for their use of AI.

A further consideration for the researchers was the asymmetrical speed of evolution between the triptych of AI-enabled capabilities. The project adopted the premise that AI accountability for policing and law enforcement essentially requires the balancing of three things: the technological (what can be done), the legal (what must/must not be done) and the societal (what people will support/expect to be done). AI-enabled technology (the possible) is changing much faster than the law (the permissible) and legislation is unable to keep pace.[7] The researchers also needed to establish the movement in societal developments (the acceptable). Looking at accountability from these three vantage points, researchers recognised that public attitudes towards the police using AI have not been captured sufficiently to know what the public expect.

Drawing on the collective experience of police officers and staff, and a survey of over 6000 people from countries across Europe, North America and beyond, the team identified the indicators of police accountability in using AI-driven technologies.

Because of the overlap in technology/law/society, it is often difficult for policing bodies to tell whether their technical specifications satisfy the law. Similarly, nailing the techno-legal issues of AI[8] doesn't tell you anything about the expectations of the citizen when you use it. To help the police and the citizen understand

[6] Chapter 4.

[7] https://hls.harvard.edu/today/is-the-law-playing-catch-up-with-ai/.

[8] For example by complying with the EU AI Act https://www.europarl.europa.eu/topics/en/article/20230601STO93804/eu-ai-act-first-regulation-on-artificial-intelligence.

how AI policies and practices will fit with the expectations or concerns of their communities, researchers sought the views of 10,114 UK citizens. Ranging from 14 to 85+years old, the majority of respondents (61.3%) across all four nations supported the use of AI in policing, *provided it came with strong accountability, oversight, and technological and legal safeguards.*

Nearly two-thirds (64%) believed that AI enhances police capabilities and the survey showed a recognition of key advantages to AI adoption in policing, especially in the context of **national security, police effectiveness** in solving crime, and freeing up **time and resources** to focus on high-priority offending.

The survey also identified some significant concerns about the police using AI, including over surveillance and the misuse of personal data. Of particular significance to biometric capabilities is that nearly 40% of respondents said they would be unwilling to contribute their personal data for the training of AI systems (only 9.3% would be very willing), and many were concerned that the police would shift 'blame' onto the technology when things go wrong. These findings are highly relevant to the use of biometrics like facial recognition technology (FRT). The survey also found regional variations in attitudes across devolved areas (Scotland and Northern Ireland), yet public space surveillance regulation remains grounded in data protection, covered by a UK-wide framework.

Crucially, the survey revealed that support for police use of AI is heavily contextual, with trust and acceptance tending to be highest where deployment carries low risks for the general public. Policing is more than enforcing laws. Although linguistically the term 'law enforcement' has been adopted and used as shorthand for policing and the police. In fact, law enforcement is a function of the police and one that many chief officers estimate to be around only 20% of policing functions and activity.[9]The survey showed the extent to which the public support AI-enabled tools for policing in the enforcement of laws is not the same as their support for it when used in non-enforcement functions such as finding vulnerable people. In accountable AI deployment, the distinction between policing and law enforcement is not mere semantics and matters practically.

The findings from the citizen survey underscore the risks of crossbreeding traditional forms of surveillance like CCTV systems with AI-driven biometric capabilities like FRT.[10] In terms of operational scenarios, the highest level of acceptance was for using AI to identify perpetrators of child sexual exploitation and terrorism, two areas that have been used elsewhere to illustrate 'perfect use

[9] See for example https://www.bbc.co.uk/news/uk-scotland-scotland-politics-38722151.

[10] Chapter 5.

cases' for remote biometrics.[11] The lowest levels **of citizen support were for the automation of core functions** like 999 emergency calls, another important finding showing that it would be a mistake to assume 'administrative' applications of AI would be less contentious than operational uses.

And finally, there is a reasonable likelihood that policing will feel obliged to use AI-enabled capabilities in the future. Developments in relation to Art 3 of the European Convention on Human Rights (ECHR) illustrate how. In a case brought by a victim of a serial rapist using a London taxicab to trap women,[12] the Supreme Court found the police liable for failing to protect citizens from degrading or inhumane treatment. The Court held that the police have a positive obligation under Art 3 to conduct an effective investigation in such cases, even where the cause of their treatment is a private individual rather than the state. The basis for liability was partly the investigative 'means' available to the police by which they might have prevented the harm. As AI-enabled preventative technology such as FRT becomes more readily available to the police, there is room for this duty to be developed by the courts. It is reasonable to expect that the police will one day have to provide a *legal defence* for their decisions not to use AI-enabled technology rather than a policy explanation.[13]

It has been recognised that AI needs a 'clear accountability process and ecosystem' (Antunes et al., 2023), while the UK government has acknowledged the need for 'clear lines of accountability across the AI life cycle'.[14] In this police-specific context, the accountability element is, by turns, a tool to provide 'transparency and human oversight of both the actions taken by the agent and the inference performed by the system' (Janjeva et al., 2023) and a method of highlighting individual/organisational accountability for misuse.

Some academics have argued that the UK government should be providing more methods to ensure greater transparency and accountability around the use of algorithmic systems in public sector decision-making (Christie, 2021). This latter function is interesting in the light of the citizens' fears that, in the future, the police would 'blame the algorithm'. As AI tools become more established and accepted in policing, this finding corroborates the importance of separating arguments that the AI technology is itself flawed or unlawful rather than the way in which the police use it. One of the grounds on which the police use of live

[11] Chapter 11.

[12] *Commissioner of Police of the Metropolis v DSD and Another* UKSC 2015/11/0166.

[13] Chapter 9.

[14] Department for Science, Innovation and Technology White Paper 2023 https://www.gov.uk/government/publications/ai-regulation-a-pro-innovation-approach/white-paper.

facial recognition technology in *Bridges* (*loc cit*) was found to have been unlawful was their failure to discharge their general, non-AI duties under the Equality Act 2010; the use of the technology was, of itself, not a breach of the duty to act 'in accordance with law'.

The full project report contains a significant amount of quantitative and qualitative data, but the accountability basics for AI-driven biometrics in law enforcement look something like this:

Lawfulness (follow the law). All aspects of AI biometrics must be lawful, and the burden of proving they are lawful sits with the police. This includes compliance with AI-specific statutes (such as the EU AI Act), formal guidance from regulators and the organisational policies of the police. Lawfulness applies in every situation. The rest are lawfulness *plus.*

Completeness (leave nothing out). AI-driven biometrics are delivered by multi-partner programmes and accountability should cover everything from procurement to deployment, including all partners and sub-contractors. Where there are gaps, the protection and promotion of fundamental rights and freedoms should prevail.

Inclusivity (leave no one out). Accountability arrangements must take account of all stakeholders engaged in and affected by the relevant AI-enabled system. Diversity should be built in and the potential for bias reduced by broad participation of stakeholders, creating policy, reviewing deployment and relentlessly looking for new vantage points.

Transparency (be open). Transparency promotes public trust and confidence, enabling those affected to make informed judgements. Accountability needs clear, meaningful, accurate and timely (CMAT) information about AI-enabled biometrics systems (subject to exceptional operational sensitivities). The police should show why using the system is necessary and proportionate, why less intrusive methods won't get the same desired result, and highlight foreseeable risks.

Impartiality (empower independence). Accountability bodies should be impartial without any material conflict of interest. External accountability—courts and regulators—will have this built in; internally, people involved in accountability for AI biometrics should at least have a degree of independence from the line management structure of those involved in their design, development and deployment.

Proof (follow the evidence). The police know how to capture, analyse and present evidence. They should approach evidence of accountability with the same rigour as operational proof, mirroring the standards of evidence gathering in terms of integrity, credibility and continuity and reflecting the impact of decisions to use available AI-enabled biometrics. Police bodies must get ready to answer for their use of AI capability and also for decisions *not to* use it in circumstances where it offers an available and legitimate tactical option in preventing and solving crime and keeping people safe.

Enforceability and Redress (make it right). Without a 'so what?' accountability is ornamental. Accountability mechanisms must give people an effective remedy, including routes for complaint and challenge, internal enforceability and redress.

Compellability (make it work). Oversight bodies must have levers to make the accountability mechanisms work. External compellability comes from legal or democratic processes, while internal policies should authorise action without the need for legal powers and litigation.

Explainability (describe, demonstrate, demystify). If you're expecting people to trust your use of AI-enabled police biometrics, you're going to have to explain what they are and how they work. Explaining the system in a technical setting is one thing; making it understood in non-technical language so that the citizen and their representatives can understand, participate and challenge its use is harder but equally important.

Constructiveness (aim for better). Accountability is more than criticism. Participating constructively with a shared aim of improvement, considering different perspectives and inviting challenge will reveal how disagreement can lead to beneficial solutions.

Conduct (hold yourself accountable). The European Code of Police Ethics states, '*the condition of a democracy can often be determined just by examining the conduct of its police*'. Police 'conduct' will increasingly include the use of AI-enabled technology, both individual and organisational, engaging professional standards, values and expected behaviours which incorporate integrity and ethics.

Learning (look for the lesson). Promote willingness to improve AI in every respect through the application of (new) knowledge and insights. When applied to the design, use *and* oversight of AI-enabled biometrics, learning includes modification and improvement of systems, structures, practices, processes, knowledge and resources, as well as the development of professional doctrine, standards and policy.

Who says the police should use AI? The question is sometimes asked rhetorically but the answer is most of the 10,114 UK citizens in the CENTRIC survey.

However, three quarters wanted *mandatory* accountability processes in place *before* the police use it. The legal analysis shows that's missing. At the same time, fewer than 20% believed there was enough accountability for police use of AI. Both findings corroborate the need for an accountability framework; both are addressed within AIPAS and the accompanying software tool to help policing bodies audit their policies and practices against its principles.

Accountability for the use of AI in policing may become a proxy measure for the extent of police legitimacy within their communities—its effective use will call for a careful balancing of the technology, the law and the expectations of society. Supreme Court Justice Lord Kerr said[15] 'A power on which there are insufficient legal constraints does not become legal simply because those who may have resort to it exercise self-restraint. It is the potential reach of the power rather than its actual use by which its legality must be judged'. The potential reach of AI-enabled biometric power in policing is almost limitless—only clear accountability will provide the constraints needed to make it acceptable.

References

Akhgar, B., Bayerl, P. S., Bailey, K., Dennis, R., Heyes, S., Lyle, A., Raven, A., Sampson, F., & Gercke, M. (2022). *AP4AI framework blueprint*. https://www.ap4ai.eu

Antunes, H. S., Freitas, P. M., Oliveira, A. L., Pereira, C. M., Vaz de Sequeira, E., Xavier, L. B. (2023). *Multidisciplinary perspectives on artificial intelligence and law*. Springer. https://doi.org/10.1007/978-3-031-41264-6

Christie, L. (2021). AI in policing and security. *Post*. https://post.parliament.uk/ai-in-policing-and-security/

Janjeva, A., Harris, A., Mercer, S., Kasprzyk, A., & Gausen, A. (2023). *The rapid rise of generative AI: Assessing risks to safety and security* (CETaS Research Reports).

[15] https://www.supremecourt.uk/cases/uksc-2013-0243.W.

11 What's (Been) Stopping the Police Using Facial Recognition?

This chapter asks the question, literally and rhetorically, why are the police not using Facial Recognition Technology (FRT) more widely. The chapter answers this frequently asked question from drive-time phone-ins, police force inspections, online conference forums and weekly interviews with journalists about how it has not gone away and provides some answers for the future.

If it's legal and it works, what's stopping them?

The police use of Facial Recognition Technology (FRT) generates a lot of material across many countries, and this particular question comes up a lot in various formats. As the UK's first Biometrics and Surveillance Camera Commissioner I would be asked a variation of this FAQ in drive-time radio phone-ins, police force inspections, online conference forums and weekly interviews with international journalists.[1] That it hasn't gone away says something about the extent to which the issues raised by AI-enabled surveillance are generally understood. The challenge deserves a response, if just for sheer persistence, but before we can get to that, the question itself needs some unpacking.

The question contains two 'ifs' and a presumption, all of which are carrying a lot of weight. The first 'if' raises the existence of a legal basis for the police to use FRT and legal authority is always a good starting point from which to examine the police using anything novel or contentious.

[1] https://www.biometricupdate.com/202105/uk-commissioner-defends-face-biometrics-use-by-police-as-practical-necessity.

F. Sampson, *How AI-Driven Surveillance is Changing our World*, Security Informatics and Law Enforcement, https://doi.org/10.1007/978-3-032-24005-7_11

In England and Wales the police certainly have statutory powers to take and retain images of people[2] in the same way as the law permits them to take fingerprints and DNA samples, along with common law powers to obtain and store information about the citizen's behaviour in public. The UK government's own Surveillance Camera Code of Practice[3] (currently on policy's death row) provides guidance to chief police officers on how to use overt surveillance systems and includes FRT. In the key legal challenge to the police use of live FRT,[4] the Court of Appeal was satisfied that the combination of statute law, regulation and police policy created a 'sufficient legal framework' covering the police use of FRT, and one that was capable of supporting its lawful deployment. However, answers to legal questions are rarely binary, and there are many finer points to cover off before the police can say their use of technology is fully in accordance with the law. For example, FRT involves the processing of biometric data, and the general data protection laws of a country will add important layers of protection for the citizen. Biometric surveillance amounts to intrusive police activity, and the law in countries like the UK will impose further obligations around covert surveillance, equality and interference with fundamental rights like freedom of speech, assembly and right to a private and home life as protected by the European Convention on Human Rights or the domestic constitution of many countries. In jurisdictions like the UK, the police will need to demonstrate that using the technology was necessary and proportionate to the intrusive impact of using it, and that the outcome could not be achieved by a less intrusive method. All surveillance activities that touch upon basic protections will be the subject of guidance from, and vigilance by, national regulatory bodies, and all must be fully taken into account if the police are to claim that their use of the technology is, in each and every case, 'lawful'.

It's important to remember that an express power to do something is not the same as an absence of specific laws preventing it and, in the context of emerging surveillance technology like FRT, we sometimes have more absence than expression. This area will need great care before any future expansion; even UK ministers have been confused about the law as Matthew Ryder (2023) evidenced in his independent review of biometrics for the Ada Lovelace Institute.

---

[2] See the Police and Criminal Evidence Act 1984, Part IV.

[3] https://www.gov.uk/government/publications/update-to-surveillance-camera-code/amended-surveillance-camera-code-of-practice-accessible-version.

[4] *R (on the application of Bridges) v Chief Constable of South Wales Police and Others* [2020] EWCA Civ 1058.

The second 'if' contained in the question relates to the technology, i.e. 'if FRT works, what's stopping the police from using it?' The use of live FRT by the police in London had a very shaky introduction[5] around 2015, when it didn't work as hoped (or required). It is fair to say that the technology deployed by the police has come on significantly since those early days. The accuracy of the technology[6] is much better, but how accurate is it to say conclusively that it now 'works'?

Whether specific biometric technology functions properly in a *technological* sense is a question that each technology partner and purchasing police force must answer for themselves—as for any other operational capability. In asking the question of the police it's worth remembering that, in AI terms, 10 years ago was the Pleistocene era: bearding police chiefs[7] with first-generation statistics is just jousting with fossils. Meaningful accountability for using state-of-the-art biometric surveillance technology requires up-to-the-minute evidence, something that will be critical for all uses of AI by the police.

But whether something is 'working' in the context we're addressing here also includes the extent to which the technology *achieves the outcomes* intended by the police in deploying it. That's a broader accountability question and links directly back to its lawfulness: if the FRT doesn't achieve what it was deployed to do, it is difficult to argue that its use was 'necessary and proportionate' to the intrusive impact of using it, or that there was no other viable alternative to achieving the same outcome.

In addressing whether the technology 'works' in policing, it's helpful to look at what sort of cases would potentially meet these requirements. Here's an example:

A man launches a terrorist attack on a crowded subway train. Detonating a smoke bomb while the carriage is between stations, he shoots ten commuters before fleeing. Finding his rental van parked nearby, along with a semi-automatic handgun, more grenades and a hatchet, the police confirm the suspect's identity. He's fired 33 rounds into a packed train and blown up multiple devices; he's armed, has several changes of clothing, and he's still in the subway system.

[5] https://www.biometricupdate.com/201509/police-use-of-facial-recognition-technology-to-be-reviewed-by-ethics-committee; https://www.biometricupdate.com/201603/uk-biometrics-report-calls-police-use-of-facial-recognition-questionable.

[6] https://www.biometricupdate.com/202407/met-police-director-says-live-facial-recognition-beats-accuracy-expectation.

[7] https://www.biometricupdate.com/202209/campaign-groups-demand-new-london-police-boss-end-facial-recognition-use-on-first-day-in-job.

Following the attack on the New York subway in 2022, I wrote an article[8] asking whether these facts presented the perfect case for the hypothetical police use of live facial recognition (LFR). The technology was *not* used on that occasion but, as the UK's Commissioner, I thought the scenario offered a compelling case for LFR. Given that the city had experienced a terrorist attack within a confined transport system equipped with an extensive camera network, with reliable images of the forensically identified suspect, if they had been able to feed Frank R. James's picture into the combined surveillance systems to find him, on what basis could the police responsibly *not* have done so? James was eventually caught but only after 36 h following a huge multi-agency search operation. He subsequently received 10 life sentences.[9]

Fast forward almost three years, and the technology had evolved. The debates and reports about police use of LFR were continuing, and another exemplar arose. On this occasion, LFR was both available and successfully deployed.

The location was Denmark Hill, London, and the facts were very different. A convicted paedophile, David Cheneler, had been placed on a police watchlist[10] because he was under a court order preventing him from having contact with children under 14 years old. In January, a police LFR camera van picked out a man walking in a street with a 6-year-old girl and matched his face with the image it had of Cheneler. Having befriended the girl and her mother—who was unaware of his previous offending or the court order—Cheneler (in possession of a knife when arrested) had collected the girl from school. The trial judge concluded there had been a sexual motivation in his actions and sentenced him to two years' imprisonment.

In answering the question 'if it's legal and it works', there are three important features for the police use of LFR in the London case. First, as the judge noted,[11] 'the technology available prevented physical contact going further'. Availability is important here, not just in terms of the equipment being accessible; it has a specific legal element too. Where the technological means to prevent inhumane or degrading treatment are reasonably available to the police, the law in England

[8] https://www.techmonitor.ai/policy/privacy-and-data-protection/facial-recognition-needs-stronger-case-law-enforcement?cf-view.

[9] https://www.justice.gov/archives/opa/pr/frank-james-sentenced-life-prison-subway-mass-shooting#:~:text=Frank%20James%2C%2062%2C%20of%20Milwaukee,%2C%20on%20April%2012%2C%202022.

[10] https://news.met.police.uk/news/sex-offender-identified-during-met-police-live-facial-recognition-operation-497544.

[11] https://www.thetimes.com/uk/crime/article/sex-offender-jailed-after-facial-recognition-spots-him-with-child-q2wbltqq5.

and Wales may not just permit the use of remote biometric technology, it may even *require* it. I'm unaware of anyone yet having relied on this human rights argument, and we won't know if these conditions would have met that threshold, but it will arise in the AI-enabled future. One thing we could be sure of back then was that the availability of AI-driven biometrics would bring a tipping point whereby the police will be challenged on their decisions *not* to use the available technology as often as they have to defend their use of it previously. That date was Tuesday 16 September 2025 when the Commissioner of the Metropolitan Police was specifically required to explain why he *refrained* from deploying LFR at a political protest organised by right-wing groups (Phillips, 2025). We can expect such challenges to increase.

The second important feature is that the person's image had been fed into the LFR cameras because he was subject to a court order and the risk of his breaching it by having contact with young children could be directly mitigated by using the technology. This was not a case of the public at large being under 'general surveillance': a court had been satisfied on the evidence presented that an order was necessary to protect the public from sexual harm by him. The subject breached that order by insinuating himself into the life of a 6-year-old girl and was found alone with her. In the Denmark Hill case we can say that the technology 'worked' in both senses—technologically because it accurately matched the offender with the image that the police had of him. And operationally because the technology allowed the police to intervene. LFR worked because it did its job.

It would be easy to celebrate the Denmark Hill outcome as a case of 'thank goodness nothing happened' to the girl but that would underestimate its significance and completely miss the legal areas where FRT will be challenged. Blanket police use of AI-enabled surveillance would probably stop something happening somewhere. But if they are to use the technology lawfully, accountably and with the support of their communities, the police must be able to demonstrate that every deployment is necessary, justified and proportionate, even when using it for slam-dunk cases like capturing a suspect during a terrorist attack or rescuing young children from the risk of sexual harm.[12] The fortuitous avoidance of remote consequences is all well and good but building trust and confidence, in the technology and the police using it, will need more than a collective sigh of relief.

[12] Which is why understanding and then avoiding the Frankenstein effect is so important—see Chap. 5.

Returning to the title of this chapter, the third part of the 'what's stopping them?' question assumes that the police have been holding back from deploying FRT. Is this a fair challenge?

When people ask what's stopping the police using FRT, they usually mean live (or real-time) facial recognition as discussed above, but there are other applications of the technology.[13] Despite its exponentially wider and deeper reach, the police take-up of *retrospective* FRT[14] has been more extensive than live matching and its use is widespread across policing in England and Wales. Live FRT, however, has had a more cautious roll-out. Why might that be? I generally use three key perspectives for answering most surveillance questions: the technological (what can be done), the legal (what must/must not be done) and the societal (what people support/expect to be done). Applying these we can see the what's stopping them question is asking: 'as it's possible and permissible, is it the *societal* element that's holding the police back from using live facial recognition?' The answer may well lie in that third perspective. Academic research shows some ambivalence in the public attitude to police using live FRT and the subject has not been fully tested in the UK, either directly by the police and their governance bodies or through parliamentary rigour. While the UK remains in the vanguard of facial recognition in policing,[15] the extent to which police use of live FRT is *acceptable to the citizen is difficult to separate from expectations about* the police use of AI generally (Akhgar et al., 2025). In a large citizen survey of public attitudes to the police use of AI generally, CENTRIC researchers found that the use cases attracting the highest level of public support were the two types set out above: terrorism and child sexual abuse. Internationally the wider picture is contradictory. In France, for example, Justice Minister Gérald Darmanin has said[16] 'If you want a secure society, you need facial recognition' yet the *impasse* in the United States[17] suggests the law enforcement technology will not become *de rigueur* for a while yet.

In the article on the Brooklyn subway case, I offered the view that facial recognition technology would be 'unsurpassable' in tackling sex offending, illegal

---

[13] For examples see Chap. 6.

[14] https://www.biometricupdate.com/202409/uk-govt-details-police-use-of-facial-recognition-during-anti-immigration-protests.

[15] https://www.biometricupdate.com/202402/the-use-of-live-frt-by-british-police-makes-the-uk-an-outlier-among-democratic-states.

[16] https://www.thetimes.com/world/europe/article/gerald-darmanin-safe-places-drug-culture-france-xzh7v09gm.

[17] https://www.biometricupdate.com/202505/new-york-senate-committee-advances-ban-on-police-use-of-biometric-surveillance.

immigration and finding individuals in breach of prison licence conditions. As police use cases go, Denmark Hill is near perfect: not so much low-hanging fruit as a root vegetable, but not all cases will be so clear-cut, and we still have some way to go to ensure all police uses will balance what's possible, what's permissible and what's acceptable. That might be what's been stopping the police; however, these use cases cover some of the most challenging policy areas, and not just in the UK. Preventing harm by those who present significant risk is at the heart of the UK government's justice reforms. What role facial recognition technology will have in those reforms remains to be seen, but it undoubtedly has something new to offer, and the UK government is not the only one exploring the potential.

Following James's arrest for the Brooklyn attack, the New York City police commissioner said,[18] 'We were able to shrink his world quickly, so he had nowhere left to turn'. AI-enabled surveillance will empower the police to shrink the worlds of suspects at a scale, pace and cost unimaginable only a few years ago; how far it should be allowed to shrink everyone else's world in doing so remains the surveillance question of our time.[19]

A couple of UK police forces were early adopters of live FRT, shouldering much of the investment risk and most of the justified blowback. The Metropolitan Police Service was one of them and their Commissioner believes, not only that the public support his use of FRT, but also that its impact could help restore their trust in policing generally.[20] The experience of his fellow early adopters in the retail sector may corroborate that belief. As trusted brand stores up and down UK high streets using FRT are seeing significant drops in crimes against their staff, customers and stock,[21] it's now the police asking what's stopping them.

References

Akhgar et al. (2025). *Survey*.

Matthew Ryder, K. C. (2023). *Independent review of biometrics for the Ada Lovelace Institute*.

Phillips, J. (2025, September 16). Met chief reveals why facial recognition was not used at Tommy Robinson protest. *The Standard*.

[18] https://abcnews.go.com/US/york-city-subway-shooting-suspect-captured/story?id=84064528.

[19] And is the subject of Chapter 20.

[20] https://www.biometricupdate.com/202309/facial-recognition-will-transform-investigative-work-says-uks-top-cop.

[21] Chapters 5 and 8.

Do Robots Have the Right to Remain Silent?

12

'You have the right to remain silent'. The inexhaustible appeal of crime-based entertainment means the English-speaking world is very familiar with the right not to say anything on being arrested. As most of that entertainment comes from the United States, it's the US version that is probably best known—the shorthand expression of 'taking the fifth' comes from the Fifth Amendment of the Constitution.

Artificial Intelligence (AI) generates lots of legal questions about how it will fit into our world—and vice versa—one of which is whether robots will have rights. The right to remain silent has, as the U.S. Supreme Court noted nearly 25 years ago, become a part of our culture and offers an insight into the rights-based arguments for AI in law enforcement.

In England and Wales, the recitation given to people before being questioned begins 'You do not have to say anything but…' and the words are from something called the 'Caution', which came into use over 100 years ago. Its younger American cousin is known in legal circles as Miranda because the constitutional right to stay quiet in the face of police questioning came from the case of *Miranda v the State of Arizona*[1] in 1966.

Arising from a case involving a suspect's confession to kidnap and rape, the right to remain silent was established as a Fifth Amendment right.[2] Miranda even became a verb in U.S. law enforcement (as in 'did you Mirandize my client officer?'). In many jurisdictions a person in custody must, prior to interrogation,

[1] 384 U.S. 436.

[2] In fact, the judgment extended a previous decision 2 years earlier *Escobedo v. Illinois*, 378 U.S. 478 (1964).

F. Sampson, *How AI-Driven Surveillance is Changing our World*, Security Informatics and Law Enforcement, https://doi.org/10.1007/978-3-032-24005-7_12

be clearly informed that they have the right to remain silent, and that anything they say can be used against them in court; they must also be clearly informed that they have the right to consult with a lawyer and to have the lawyer with them during interrogation. Failing to remind suspects of their rights properly can lead to evidence being excluded and cases being thrown out—many cases still are.

Lawyers know it as the right against self-incrimination; expensive ones call it by the Latin maxim *nemo tenetur seipsum accusare*[3]—*either way* it's a sacrosanct feature of the criminal justice system. As more decisions and their consequences begin to flow from AI-enabled devices, one question for the future of surveillance is whether robots have the right to stay silent?

For self-incrimination to occur, you need two things: a crime (where the word comes from) and a risk of conviction (where the warning comes from). With a robot, there's currently neither. A robot can't testify against anyone because that function is usually reserved to 'live' witnesses,[4] nor can it be the subject of a criminal conviction. In shorthand, 'Taking the Fifth' (remaining tactically silent) gives citizens several crucial civil rights as against the state, and similar provisions exist in the UK. Without rights, a robot can't take the Fifth as it has nothing in jeopardy when it speaks. That seems to deal with that then: asked and answered. No rights. No risk. No further questions.

But should there be? Consider three scenarios. In the first, a court needs access to social media posts during a prosecution. The data is necessary for the purposes of the trial and the court orders disclosure by the people holding it. If the data is held by another defendant in the same case and is likely to incriminate them, the judge will have to balance a few tricky issues, weighing them against each other, and the data may be withheld. But what if the data is held by an AI robot? Can the robot be forced to give it up? If we look at this, not from the robot's perspective, but rather from that of the human whose constitutional rights (to a fair trial etc.) are in peril, the answer seems easy. The AI should not be permitted to 'stay quiet' where the established rights of the defendant are fundamentally undermined by it doing so. The civil rights *of the human* on trial extend to any relevant data held by the AI, in which case it should not be permitted to remain mute. This at least fits with the norms of 'ethical and explainable' AI.[5]

Some might say the robot in this situation is in fact under a *duty* to contribute any relevant data to the trial process. The problem with this is if it's inappropriate to speak of rights for machines, then they shouldn't have duties either. Rights

---

[3] No one is obliged to incriminate themselves.

[4] Or occasionally statements produced by someone now dead.

[5] Chapters 1 and 10.

and duties are reciprocal and are a good example of how the traditional labels of criminal justice won't really fit the future AI-enabled criminal justice world. However, robots will generally be owned and operated by somebody—where this is a private company, there could be further challenges in accessing its thoughts.

In answering the question 'do robots have the right to remain silent?' it's important to keep in mind why any of us enjoy that right and where it came from. The feeling of compulsion to share information with the police and to answer their questions is well documented throughout criminal jurisprudence. As the judgement in Miranda said, '*unless adequate protective devices are employed to dispel the compulsion inherent in custodial surroundings, no statement obtained from the defendant can truly be the product of his free choice*'. In Miranda, Chief Justice Warren reviewed the tactics of police interrogation and the pressures of finding oneself under investigation. Can those pressures be felt by an insentient robot? Probably not (yet) but the realities of the criminal justice machinery should be borne in mind by everyone building or buying one.

Robots will come with a new set of challenges for identifying the existence and exercise of 'freedom of choice'. Freedom to choose brings freedom to be wrong and there are many examples of the computer being catastrophically wrong in the criminal justice setting, as AI decision-making proliferates, it's reasonable to anticipate many more.

Consider the second scenario. In this one the robot in question is a policing tool running the national Automated Number Plate Recognition system (ANPR). Monitoring strategic movement of sensitive traffic around the country, the robot makes an error and wrongly identifies the licence plate of a vehicle as transporting nuclear material to one of many new Small Modular Reactors (SMR) recently built as part of the national commitment to clean energy. The vehicle deviates from its filed route and the AI automatically alerts the police who, properly relying on the AI's information, locate the vehicle and try to bring it to a halt. The driver does not stop as requested, whereupon the officers, reasonably and objectively believing that the situation represents an immediate threat to life, force the vehicle into a 'hard stop'. The driver is fatally injured during the manoeuvre. It transpires that the AI wrongly flagged the vehicle based on one of 160,000 number plate misreads that occur every hour; there was no nuclear material on board and the driver had simply panicked when the police blocked him in. At no time before the police intervention did the driver or the vehicle represent a risk to anyone. The AI is in possession of all the information necessary to understand what happened and to provide a public explanation. In such a scenario could there be any legitimate reason for the police robot to 'remain silent'? How else would the officers defend themselves legally as is their right in any fair prosecution?

Now reverse the error and look at the scenario where the robot fails to identify a vehicle that *is* in fact transporting radioactive material and that has been hijacked. There is no need to elaborate this scene any further to appreciate its impact, not just in the relevant country, but in every one of the 80 countries that have now signed up to building SMRs in their neighbourhoods (Sampson & McNeill, 2025). On these facts alone, public confidence in the resurgence of nuclear energy would be damaged and a raft of mitigating measures would be demanded. A human operator could take the Fifth but can the robot stay quiet?

Drawing a bright line between human and technological immunities will not be easy but it will be necessary and there are legal ramifications in doing so. We might accept that robots themselves have no individual basis for withholding data in the justice process, but it may be in the interests of their human designers for their progeny to keep schtum when interrogated after a serious error. Given that AI systems *inform and support* critical decisions, as opposed to making them, the lines separating culpability for the processing of data and decisions taken in reliance on it are going to be crucial. If the robot is to be treated as a proxy for the malfeasance (or misfeasance) of its overseers, those who design and deploy it might argue that they have a right to remain silent during criminal investigation into the same issue, and that making their robot produce evidence against them is arguably an indirect breach of that right. There is also the issue of communication. An integral part of the right to remain silent and to have access to legal counsel is the right not to be held incommunicado, but that is an entirely human right, along with regulated conditions of detention. What is the status of a robot in police custody—stored property or detainee? Access to robots will be critical to prosecutors and defence teams alike. If they are kept powered up and functioning, how will that work? Are robots to be treated as technological devices and exhibits, locked in Faraday cells, sealed off from the outside world? Without doing this, the integrity of any data they have to share will be fatally compromised.

These are all questions of policy, but more immediate practical problems also arise. It is easy to see how, for example, programmers might build in their own 'Heaney switch', guaranteeing the robot's silence by making it *incapable* of self-incrimination, thereby achieving by coding what our courts have identified as a cherished constitutional right in humans—and programming in Seamus Heaney's famous admonition: 'whatever you say, say nothing'. Alternatively, robots could be programmed to reflect their Sixth Amendment rights, the right to legal counsel, and go into protective mode when being compelled to disgorge evidence. Such a function could have a legal advice filter identifying matters that are 'admitted', matters that are 'not admitted' and matters that are 'denied'. Thus programmed, rather than staying mute, the robot selectively redacts data that might attract any blame coming the way of them or their maker, and gives up any data that it deems harmless. This is what humans with lawyers do.

An alternative strategy might be to design the robot to deliberately blurt out *disinformation* and leave false evidence trails, creating Deepfakes and scattering digital chaff. In a legal system that doesn't allow a robot to give evidence, it would not be in the frame for perjury. Who in such a scenario would be lying?

One way to address the problems would be for legislation requiring 'candour by design' to be built into all AI used for critical functions like policing, in much the same way as privacy by design approaches required by regulators.[6] Far from allowing robots to take the Fifth, this would mean they must be designed with a 'veracity button', a kind of AI flight recorder. In policy terms this would fit with the political *zeitgeist* in the UK where the government believes an act of parliament is needed to make the police and other public servants tell all following a critical incident.[7] But imposing a 'duty' of candour on AI designers brings some practical challenges, one being the risk of programmed sword-falling. Mandated admission may lead to a Spartacus factor where all the robots involved claim to be the offender, not just to show compliance with the candour code, but specifically to exonerate any humans in the chain of culpability.

At this point Netflix taggers and armchair gumshoes will detect another hackneyed cop drama episode, the one where a hapless stooge is persuaded to take the fall for everything and the plainly guilty walk. Could Patsy-AI be programmed to 'fess up simply to divert from the humans in—or out of—the loop? In a recent national survey of over 10, 000 UK citizens, many expressed concern that, in the future, the police would shift the 'blame' to their AI tools when things go wrong.[8] Accountable use of AI-enabled capability requires transparency by the police.[9] However, there will inevitably be times where the public interest depends on the robot *not* giving out information: national security is the obvious one and might apply to our second scenario above. History has shown how humans, trained and deployed to exercise discretion and restraint in this area of operations, have proved to be lamentably imperfect. While not directly involving the use of technology, more extensive and enduring intrusions such as undercover

---

[6] See e.g. https://ico.org.uk/for-organisations/law-enforcement/guide-to-le-processing/accountability-and-governance/data-protection-by-design-and-by-default/.

[7] See the Public Office (Accountability) Bill—https://bills.parliament.uk/bills/4019.

[8] Chapter 10.

[9] What this looks like is discussed in Chapter 10.

officers deceiving people into having intimate relationships and even starting families under their confected identities to gather intelligence have seriously damaged public trust in such covert operations[10]—will using covert surveillance robots be any better?

The overarching challenge in the second scenario is how we will ensure our AI-enabled police surveillance tools are any better at knowing when to be candid or quiet than we have been thus far. Simultaneously designing police AI for transparency and public interest immunity will be challenging.

And so to our final scenario. A penitent crime boss reflecting on his misspent life decides to confess all, not to the police, but in a confession box just installed in Peter's Chapel, Lucerne. The confessional is not occupied by a human priest, as happened in the past. The church having embraced AI-enabled technology, the job of hearing confessions has been given to an 'AI Jesus' in residence.[11] The contrite crime boss sits on the other side of the curtain, revealing his many crimes to an algorithm.[12] Investigators have been keeping him under surveillance and want to get at the gospel truth, the invaluable information which includes details of murders in several countries and the disposal of victims' bodies, crimes that have remained unsolved for decades that have left families having no information about the fate or whereabouts of loved ones. Is the Fifth Amendment engaged? Can the Feds compel AI Jesus to take a 'queen for the day' proffer agreement? However spiritual the circumstances, the only legal sanctity for the communication between the confessing and the robot here is contractual. Unless the courts grant some special constitutional status to inculpatory algorithms, we know how this probably ends. Litigated in the pre-AI era, the long-running saga of the Boston College tapes is insightful (Sampson, 2016). Courts in the US and the UK decided that digital recordings of admissions made by members of terrorist organisations in the course of academic research were not protected by confidentiality agreements. In our scenario, the law enforcement agencies on both sides of the Atlantic will probably get the court's blessing—if not the church's—when they go after confessions for serious crimes such as terrorism, regardless of any data sharing niceties. We should expect the same approach in respect of robots. Any attempt to suppress confessions made to an AI is likely to be met with the same legal arguments in favour of the public interest.

---

[10] *Kate Wilson v (1) Commissioner of Police of the Metropolis (2) National Police Chiefs' Council* IPT/11/167/H—https://post.parliament.uk/public-trust-and-confidence-in-the-police/.

[11] https://www.theguardian.com/technology/2024/nov/21/deus-in-machina-swiss-church-installs-ai-powered-jesus.

[12] Now there's an idea for a plot.

Not all cases, however, will involve terrorism and murders; as our use of AI-enabled technology becomes more routine, we can expect routine law enforcement applications for access to our devices *and any instructions that we gave them* to increase.[13]

Many core elements of an effective criminal justice system like a fair trial and independent legal advice depend on the availability of reliable information. Regardless of whether they have any rights, properly programmed bots will know who has; they will know the law—including all authorities and judgements—and will not need to be Mirandised because they will know that anything said may be used against others.[14] They shouldn't even need an attorney.[15]

AI-enabled tools are already bringing transformational changes to policing and law enforcement, examples of which are discussed throughout this book. One of the many questions that Agentic AI will bring is its potential contribution to the criminal justice system. Will robots be allowed to testify as witnesses? With their unsurpassed ability to recall, compare and—if directed by a judge—ignore information, shouldn't robots be allowed to sit on, or at least alongside juries? They will remain attentive and objective but is that all we want from our justice system—bloodless processors of data? Not according to some research which found a combination of AI and judicial decision-making is preferable.[16]

Questions about the impact of AI-enabled technology in policing are getting closer; for some, the answers may already be here. There are still daily disputes in the courts as to whether the police told the suspect of their rights and, if so, whether the correct words were used in a way that the suspect understood. With that in mind, shouldn't the police at least be using an AI assistant to make sure suspects get cautioned properly, at the right time, auditably and accurately, in any language necessary? 'Hey Miranda, read them their rights'.

[13] Thomas Brewster 'DHD ordered OpenAI to share user data in first known warrant for ChatGPT prompts'. Forbes, October 20, 2025.

[14] Interestingly, the Supreme Court in Miranda said this didn't matter (for humans): "The Fifth Amendment privilege is so fundamental to our system of constitutional rule, and the expedient of giving an adequate warning as to the availability of the privilege so simple, we will not pause to inquire in individual cases whether the defendant was aware of his rights without a warning being given… whatever the background of the person interrogated, a warning at the time of the interrogation is indispensable to overcome its pressures and to insure that the individual knows he is free to exercise the privilege at that point in time" [469].

[15] Danny Fortson, Luminance boss: lawyers can survive AI, but I'm not sure how many. The Times 19 July, 2025.

[16] Nikki Rojas, Does AI help humans make better decisions? The Harvard Gazette, June 14 2024.

References

Sampson, F. (2016). Whatever you say…the case of the Boston college tapes and how confidentiality agreements cannot put relevant data beyond the reach of criminal investigation. *Policing: A Journal of Policy and Practice, 10*(3), 222–231.

Sampson, F., & McNeill, J. (2025, January 13). *National security, local Jeopardy*. Centre for Emergency Technology and Security, Alan Turing Institute.

Is This the World's Most Unreliable State Surveillance System? 13

Few things have driven biometric technology like identity. The imperatives to secure financial, digital and physical borders have supercharged the global emergence of biometric ID solutions. One glance at the daily *Biometric Update* news website will show the pace and breadth of the rush to establish national identity structures, for example, Malaysia's target of 15 million users for its national digital ID by the end of 2025 and Ethiopia's of 90 million by 2027.[1]

If you're a human, the technology sector is introducing new levels of security and ID capabilities for individuals every day, while FIDO's user authentication standards[2] offer 'public key cryptography' techniques to provide user-friendly, phishing-resistant authentication, so people can drop passwords which are notoriously hackable. If you're a company, there's a Global Legal Entity Identifier Foundation (GLEIF) founded on the basis that each business worldwide should have only one global identity and that identity should include a digital identity.[3]

But there's one critical outlier, a dependency covering all sectors of our national infrastructure that has seen no meaningful technological improvement in decades and now represents perhaps the greatest ID-based vulnerability worldwide. While innovations like multi-level authentication (MLA), age confirmation and liveness detection have powered national ID capabilities for citizens, the system for reliably identifying our vehicles has remained stationary.

---

[1] https://www.biometricupdate.com/202510/biometrics-get-ai-market-lift-amid-acquisitions-national-id-ambitions.

[2] https://fidoalliance.org/specifications/.

[3] https://www.gleif.org/en.

F. Sampson, *How AI-Driven Surveillance is Changing our World*, Security Informatics and Law Enforcement, https://doi.org/10.1007/978-3-032-24005-7_13

National investment in mobile drivers' licences (mDL) has involved multiple millions of dollars (and Yuan, Yen and Euros) as the countries recognise the importance of having reliable ID capabilities for vehicle drivers and road users. The increase in verifiable digital credentials (VDCs) will allow AI-enabled capabilities to verify mDLs using secure connections between AI agents and applications. Some companies are pen-testing the technology, working out the practicalities of combatting fraud and cyber-vulnerabilities.[4]

From a strategic surveillance perspective, however, when it comes to driver vs vehicle identification, it's the latter that matters most. Macro management of traffic-dependent risks like health pandemics and environmental hazards relies almost entirely on accurate vehicle identification. Other aspects of critical infrastructure protection, resilience and monitoring, like supply chain continuity, civil contingencies and national preparedness, depend heavily on the state's ability to know what vehicles are going where and when. Yet, despite the strategic importance of this ID data, the UK and almost every other country on Earth rely on a physical credential stuck on the outside of a vehicle by its user.

Hailing from an era of physical roadside inspection by a constable, the licence plate (or number plate as it's known in the UK) is an anachronistic feature of the analogue surveillance age. When formal identification of a vehicle consisted of visual human reads, a slightly obscured plate—or no plate at all—could be spotted, investigated and rectified. But strategic roads management data is now collected and processed by networked cameras and their ability to detect errors and inconsistencies is woefully poor. Recording as many hits-per-second as Instagram, the UK's Automated Number Plate Recognition (ANPR) system is a formidable tool but in identifying the country's 41.2 million registered vehicles[5] it has a single and failing point of weakness: what my predecessor called the 'humble number plate'.

The licence plate is a vehicle's national ID card—one that you can order online or over the counter from any of almost 45,000 registered suppliers (that's nearly one for every 1000 UK drivers), all using uncontrolled stationery and ineffectual manufacturing standards. Enforcement is supplied by a network of cameras which are unable to detect flaws, forgeries or fakes. If they can't read the plate at all they simply let you pass unrecorded.

[4] https://www.biometricupdate.com/202410/mdl-verification-moving-from-theoretical-to-practical-questions.

[5] https://www.gov.uk/government/statistics/vehicle-licensing-statistics-april-to-june-2023/vehicle-licensing-statistics-april-to-june-2023.

Researchers at Cranfield University were involved in setting up the UK's ANPR system decades ago and have a great deal of policing and technological experience in this field. They have long known that registration is no guarantee of even basic compliance and have used examples of cases where criminal networks are exploiting the yawning vulnerabilities in the surveillance system.[6] They have been able to demonstrate the inadequacy of the Government's registration scheme for plate sellers by approaching properly registered members of the scheme and asking if they were willing to sell them number plates specifically to defeat ANPR or emission zone cameras. The researchers, some of whom previously worked for the police, point out that, under the current scheme, registered suppliers in the UK can sell you a number plate with any alpha-numeric combination you like, provided two conditions are met. The first condition is that they call it a 'show plate'; the second is that you must promise not to use it on a road. In comparison with other regulatory standards for vehicles, putting our trust in this level of 'authentication' and 'supply control' is bewildering—(would we do that with tyres or brakes?); in a digital age where awareness of, and investment against, ID fraud and scamming, it is also breathtakingly naïve.

The Government's registration scheme for non-show plates is little better and relies on registered suppliers demanding proof of ID from the buyer, who must also show their entitlement to display the number they're buying—researchers found many registered suppliers who didn't ask for either.[7] In fact, of the 200 registered suppliers that they have approached so far, not one has asked the researchers for relevant proof of ID or asserted any other 'security controls'. Admittedly, that still leaves another 44,800 who might but the odds in favour of strict regulation aren't good.

As dependence on traffic cameras expands, incentives to game the vehicle ID system increase. Using an inherently unreliable ID system to enforce financial measures like the collection of vehicle tax and emission zone penalties is where the system's absurdity becomes most acute. In September 2023, former Surveillance Camera Commissioner for England and Wales, Tony Porter, estimated that two million motorists could be doctoring number plates to evade Ultra-Low Emission Zone (ULEZ) style cameras,[8] and Transport for London

---

[6] Man made thousands of number plates for criminals who carried out multi-million crimes around the West Midlands including keyless car theft https://www.bbc.co.uk/news/articles/cw44m7xjwpwo.

[7] https://www.theguardian.com/technology/2025/jun/02/cloned-number-plates-investigation.

[8] https://www.telegraph.co.uk/news/2023/09/06/ulez-speed-cameras-number-plates-doctored-by-drivers/.

reported an eightfold increase in the number of penalty charge notices (PCNs) issued between 2019 and 2024, which had to be cancelled because the plates displayed on the offending vehicle had been cloned.[9]

Tens of thousands of drivers are receiving substantial fines and penalties because their vehicle's 'identity' has been copied, and the volume is accelerating[10] at the same time as the UK's new 'Data Act' promised secure digital identities to improve efficiencies, create positive customer experiences and reduce fraud[11]; someone can still clone your vehicle as easily as buying a pair of driving gloves.[12]

Among the challenges the Group have faced in getting their message across the Group have been faced with technical complexity and the inability to gather evidence, but neither has discouraged them. Yes, understanding the impact of specular plastic surfaces on reflectance conformity with reg 11 of the Road Vehicles (Display of Registration Marks) Regulations 2001 (SI 2001/561) (as amended) isn't clickbait even for the ID community, but every other sector (finance, travel, security, telecoms, health) has found a way through similar tedium. As for the inability to collate hard proof—e.g. "if you don't know how many tax pounds have been lost since the Government scrapped the paper tax disc how do you know it's a problem?[13]"—every forensic examiner and research student will tell you that, when searching for proof, the absence of evidence isn't evidence of absence. When you're dealing with numbers and consequences of this size, the lack of data should be a hazard warning light.

In their submission to a review by the PACTS, they pointed out several significant shortcomings in the UK vehicle ID scheme, including:

- There is no agreed specification for a registration plate, nor definitive or representative diagram depicting the requirements of a registration plate.
- As the 'customer' for registration plates, the Government has neither provided nor stated its requirements.
- The law governing the display of plates is ambiguous, contains anomalies and is open to variable interpretation, predictably resulting in variable registration

[9] https://tfl.gov.uk/corporate/transparency/freedom-of-information/foi-request-detail?referenceId=FOI-0616-2526.

[10] https://www.thetimes.com/business-money/money/article/criminals-cloned-my-car-so-I-got-fined-v73z85r8b.

[11] https://www.yoti.com/blog/uk-data-act/.

[12] https://www.bbc.co.uk/iplayer/episode/m002fmkl/morning-live-series-7-08072025.

[13] This is a genuine question that some officials have raised.

plates, with the relevant British Standard (BS AU 145d) containing misleading diagrams and the revision (145e) falling short of listing the properties required of registration plates.
- The Government permits variability of design size and other properties.
- Registration plate abuse is widespread with members of the Government's scheme being prepared to make cloned and false plates.
- Increasing numbers of innocent citizens are being prosecuted and harassed for fines and infringements they did not commit.
- The Treasury is unable to quantify how much vehicle excise duty is being lost by relying on the plate/ANPR system.

The UK's vehicle ID system relies entirely on the compliant display of a corruptible and fungible credential, enforced principally by a camera system that can be defeated by one of many minor variations open to the user. Imagine if that were the case with, say, passports or employee access controls. As an illustration, by simply placing a 3 mm plastic film over the retroreflective part of the plate (decreasing its reflectance and making the plates darker by 30–40%), you can make the numbers and letters unreadable by ANPR cameras at night but in a way that is undetectable to the naked eye in daylight.

To depend on such a system for strategic traffic management and domestic infrastructure is almost risible and it is why I reported to Parliament[14] that ANPR has itself become part of our Critical National Infrastructure. Quietly and unobtrusively, ANPR supports pretty well all the functions set out in the government's just-published National Security Strategy.[15]

Sophisticated customer ID procedures, supplier management processes and anti-money laundering protocols have become a way of life for all of us. Rapid advances in remote ID and authentication in areas like banking, online behaviour and even basic household security highlight the asymmetry in the regulatory approach to vehicle identification. We have recognised that the principal surveillance risks in virtually every other setting of our lives come from the technology being corrupted or hacked.[16] In contrast, the national vehicle ID system doesn't need Deepfakes or impersonation software to fool it and there's no need for deep

[14] https://assets.publishing.service.gov.uk/media/63e3ce2fd3bf7f17347092c4/Biometrics__Surveillance_Camera_Commissioner_Annual_Report_21-22.pdf.

[15] https://www.gov.uk/government/publications/national-security-strategy-2025-security-for-the-british-people-in-A-dangerous-world/national-security-strategy-2025-security-for-the-british-people-in-A-dangerous-world-html.

[16] Chapter 2.

packet inspections (DPI) or fiendish cyber tools to expose its many weaknesses: they're right under our noses every time we take to the road.

And this situation is not unique to the UK. While some in-vehicle variants such as radio frequency ID (RFID) tags are being tested in some countries, they are equally vulnerable to cloning and corruption as number plates.

Even where a vehicle *is* displaying properly made and correctly allocated plates, ANPR cameras have misreads. In the UK, the police have historically set an 'acceptable' misread level at 5%. With daily reads approaching 100 million, 'acceptability' means over 200,000 mph (misreads per hour). Viewed through a cyber-security lens, that's an unfathomable number of common vulnerabilities and exposures (CVE) for a system that's supposedly protecting our critical infrastructure.

In the age of increasingly intrusive surveillance capability, our vehicles enjoy far greater privacy than we do. That may be why ANPR vulnerability has attracted so little attention, but in the hot debate over police ID technology, the ratio of attention spent on Facial Recognition vs Vehicle Recognition is surprising. This imbalance will become more conspicuous and more important as autonomous vehicles begin to appear on our roads.

Recently, the revenant issue of the 'ghost' plate has been picked up again by MPs[17] who have provided an opportunity for some of us to make the case once again for vehicle ID, but we are so far behind. As a UK parliamentary group convened to consider improvements to the number plate, one company was releasing a beta version of its decentralised identity framework, introducing post-quantum cryptography to protect digital credentials of the individual.[18]

When considering the arrangements for vehicle ID, I propose starting with one simple question[19]: is the framework giving the public what they need? For vehicle licence plates the answer is no; the evidence is abundant and the search for an alternative is pressing. Formal reports to ministers by the last two Surveillance Camera Commissioners on ANPR's strategic vulnerabilities, however, have produced little beyond boilerplate rehearsal of assurances. When they are debating controversial new police powers to stop protestors hiding their identity,[20] UK

[17] The Cranfield Vehicle ID Group's response can be seen https://www.cranfield.ac.uk/centres/centre-for-computational-engineering-sciences/vehicle-identification-grouphere.

[18] https://www.nature.com/articles/s41598-025-16315-8.

[19] The same question applies to any regulatory framework debate.

[20] The Crime and Policing Bill—https://bills.parliament.uk/bills/3938.

lawmakers might note that the protestors' ability to conceal the ID of their vehicles with impunity will remain as easy as pulling a woollen balaclava over their faces.

The vehicle registration plate is an ID relic from another era. In the UK it's governed by an underpowered regulatory framework that insists it's keeping up despite the lap time evidence to the contrary. Looking down the biometric highway, the latest Police Emerging Science and Technology Trends report[21] sees UK policing one day using brainwaves to confirm individual ID. In comparison, the system of vehicle ID looks like a horse-drawn solution to an F1 problem; our continued reliance on it leaves us profoundly exposed, to the extent that it arguably meets the definition of a major system failure in the UK's National Risk Register.[22]

Alternative technologies will need careful, informed research but cracking the challenge of a remotely-readable unique vehicle identifier (RUVI) will bring a once-in-a-generation, globally scalable market opportunity. The Marshall's green flag is up and the grid's clear, so what's holding us back?

---

[21] https://science.police.uk/delivery/resources/police-emerging-science-and-technology-trends-report/.

[22] https://assets.publishing.service.gov.uk/media/67b5f85732b2aab18314bbe4/National_Risk_Register_2025.pdf.

14 Policing Deepfakes—Can the Camera Ever Lie?

There's a well-known saying that 'the camera never lies'.

Of course, even in its early days, photography wasn't a direct reflection of reality, and the saying wasn't meant as a literal axiom. Many features about a photograph, like the choice of subject, aspect, timing, lighting and other variables, were largely in the subjective hands of the photographer. Processing and printing introduced further opportunities for intervention between the moment and the record of its capture; some artists physically altered the image during the chemical development process.[1]

Often used when something or someone is recorded in a specific situation and/or at a given moment, the expression implies that the very existence of a picture is undeniable proof of the event. This probably has less to do with trusting the image and more about the fact that old-fangled cameras were mechanical devices. The box brownie was *incapable* of lying, ergo whatever it was pointed at and its photosensitive film exposed to must (however artistically presented) represent a basic truth (e.g. that you met this person, visited that place, wore a hat, carried a guitar, etc.).

Being able to trust the image is a critical feature of police surveillance cameras.

In a criminal justice setting, if the police are going to rely on a photograph as evidence or for intelligence purposes, it is vital that they can prove its provenance. This is why forensic teams obsess about the evidential integrity of pictures and the undisturbed scenes at which they were aimed. In the evolutionary cycle of technology, the surveillance device of the first era was a passive tool of capture

[1] https://www.rauschenbergfoundation.org/art/galleries/series/bleacher-series-1988-91.

F. Sampson, *How AI-Driven Surveillance is Changing our World*, Security Informatics and Law Enforcement, https://doi.org/10.1007/978-3-032-24005-7_14

in the hands of an active operator[2] so, for the criminal justice system, this was a mechanical issue. Once digital photography brought us the ability to manipulate our pictures however, offering a dazzling palette of options to insert, remove and edit the entirety of the image, we left any idea of 'cameras' behind.[3] In the second era—the one we are just leaving—the box with chemical coated film became a computer capable of many functions, a basic one being to record images. Able to follow instructions and carry out complex functions, using edge computing and cloud storage to allow faster processing of metadata, the digital surveillance device replaced Old World primates like CCTV and opened the gates for AI-enabled remote biometrics.

Unfortunately, the laws and principles of surveillance did not evolve alongside the technology and the existing rules of evidence and procedure continue to apply largely unaltered.[4] To be used as evidence, digital images and records need the same established level of integrity and verification as any other forms, and the internet, social media and cyber-enabled criminality have brought significant new challenges to criminal investigations. Early challenges for the criminal justice system arose in the reliability and fairness of using material from social media accounts, the weight given to self-serving posts and unauthenticated video feeds (Akhgar et al., 2014). But nothing so far has brought the change and challenge of Deepfakes.

Imagine you are a juror sitting on a prosecution for a murder alleged to have happened during a birthday celebration. Images from multiple devices clearly show the defendant at a party in a house around the time of the killing, supported by witness statements from others who also appear in the pictures. The images are corroborated by social media posts, along with a receipt for a local fuel purchase made by the defendant and further time-and-date-stamped shots of him with his vehicle on a forecourt security camera which are consistent with records obtained from the police Automated Number Plate Recognition (ANPR) system.

Nothing in this scenario would be exceptional—unless, that is, the digital evidence has been submitted, not by the prosecution, but by the defendant as his Deepfaked alibi to show he was at a party several hundred miles away from the scene at the time. How would the prosecution show that all the images were

---

[2] For a closer look at the evolution of surveillance tools and our relationship with them see Chapter 1.

[3] The *camera obscura* concept involves the capture of light through an aperture into a dark container, leaving an image from a source in front of its lens.

[4] Raising the title question 'Do robots have the right to remain silent?'—and discussed specifically in Chap. 12.

Deepfakes and the records digital frauds, particularly where the police ANPR reads support them?

The prosecution may also have images, social media posts and digital data contradicting what the suspect and his witnesses are saying, but how are jurors to know which ones are the most reliable? Using inherent vulnerabilities in legacy systems like ANPR[5] (in this case, simply by getting someone else to drive around in a similar vehicle bearing the same number plate) to corroborate Deepfakes—and vice versa—will make it challenging to undermine them. In a criminal trial there is usually an imbalance in who must prove what, with the majority of the burden falling (properly) on the state. If the police are relying on images as evidence, it is vital that they can prove the provenance and leave no reasonable doubt, whereas someone providing an alibi only has to show theirs is likely to be true. And that's what Deepfakes are very good at.

Image synthesis has potent appeal. As entertainment, the ingenuity of Deepfakery has been contagious, and overtly faked photos of celebrities ('Pope in a puffer', Donald Trump wrestling with New York's finest) have become legendary. Deepfakes are irresistible to fraudsters, and the UK government has said[6] their sheer scale, combined with their greater sophistication and convincingness, makes Deepfakes 'arguably the greatest challenge of the online age'.

Deepfakes have not just accelerated the pace of change in digital surveillance; they have fundamentally altered its direction and triggered the next era in its evolution.

Part of the UK government's response was the establishment of the Accelerated Capability Environment (ACE), maturing AI content detection tools like EVITA (Evaluating Video, Text and Audio) and shifting the focus away from volume. With daily attempts to outdo each other for originality and undetectability, creators are astounding the online consumer with visual prestidigitation, stimulating experimentation and offering technical support.

In assessing the challenges this brings for policing, the government identified the biggest challenge as being in digital forensics, where officers can be faced with up to a million child abuse images on a single seized phone. In tackling those challenges, the government has focused on raising public awareness, identifying technical standards and deploying AI tools to detect the use of other AI tools. The ACE team ran a Deepfake Detection Challenge to create a reusable

[5] https://www.biometricupdate.com/202507/is-this-the-worlds-most-vulnerable-national-id-system.

[6] https://www.gov.uk/government/case-studies/innovating-to-detect-deepfakes-and-protect-the-public.

'gold standard' dataset designed to test detection models, including those targeting child sexual abuse material (CSAM). There are, however, other forensic and surveillance scenarios where the risk from Deepfakes is great and the challenge for policing comes from sophistication and scale.

In the Deepfake era there are several surveillance-related risks to policing and the criminal justice system on the horizon, and we haven't yet planned for their arrival.

In the UK[7] the defendant is allowed to adduce evidence tending to show that, by reason of their presence at a particular place or in a particular area at a particular time, they were not (or were simply *unlikely* to have been) at the place where the offence is alleged to have been committed at the time of its alleged commission.

The above scenario is, of course, speculative and would clearly require a level of organisation and deviousness by several people (although such people and practices are not unknown to the criminal justice system). But with an almost infinite range of potential criminal uses for Deepfakes, the growth in the Deepfakes-as-a-service market has exploded.[8] One future use will probably be related to the provision of digital evidence to support alibis. You don't need any technical expertise or aptitude in order to use Deepfakes for your criminal enterprise—in an online service economy you can order Deepfakes like any other transaction. As courts begin to see the first—albeit crude—examples of Deepfaked evidence,[9] the ready availability of undetectable imagery, fraudulent digital records like receipts[10] and the intrinsic vulnerabilities of vehicle identification systems,[11] will make it very challenging for the police to disprove some digital evidence and doing so will be resource-intensive—the more that defendants try to evade justice by using Deepfakery, the more resources the police will need to counter it.

A second, much bigger, challenge from Deepfakes doesn't even require an 'as-a-service' provision and comes from our behaviour and the changing nature of surveillance itself.

---

[7] See the Criminal Procedure and Investigations Act 1996.

[8] https://www.datensicherheit.de/wp-content/uploads/tehtris-deepfake-as-A-service-threat-intelligence-report.pdf.

[9] *Mendones v Cushman & Wakefield Inc et al.*, case no.23CV028772. https://www.ddg.fr/actualite/when-deepfakes-enter-the-courtroom-a-landmark-california-decision.

[10] https://www.computerworld.com/article/4079701/new-ai-scam-fake-expense-receipts.html.

[11] Chapter 13.

Policing is beginning to rely on images uploaded and shared by the citizen, something that is changing the surveillance relationship between us and the state.[12] Ignoring citizen uploads carries substantial risk for the police but, if they are to rely on them or discard them, either way the police must verify the material's authenticity—and that is becoming much harder.

With Deepfake fraud attempts increasing by 3000% in 2023 and 60% of consumers encountering them in the past year,[13] World noted that 'every call and video now carries doubt.'[14] Some of those calls and videos will be shared with the police as evidence of crime or a need for intervention and response.

Biometric Update reported recently,[15] 'videos made with OpenAI's Sora app are flooding TikTok, Instagram Reels and other platforms, making people increasingly familiar—and fed up—with nearly unavoidable synthetic footage'.[16]

Some online content that we take as true, sharing with others and posting comments, may have been created by state-directed Deepfakes with the intention (or effect) of causing discord, disbelief or disorder. Even seasoned journalists are hoodwinked by faked interviews with political figures[17] and we have seen how AI 'bots' have been used to flood national application processes for things like student financial aid, political defections[18] and war zone reporting,[19] causing some observers to conclude 'even public sentiment can't be trusted'.[20]

---

[12] Chapters 4 and 5.

[13] https://www.eftsure.com/statistics/deepfake-statistics/.

[14] https://world.org/blog/world/the-rising-tide-of-digital-fraud.

[15] https://www.biometricupdate.com/202510/tale-of-two-platforms-sees-sora-flood-web-with-deepfakes-world-save-the-day.

[16] At the time of writing, I can get a Sora AI video engine by downloading ChatGPT 5.0. Within five days of the latest version being released, one million people did just that. https://www.bbc.com/news/articles/crkjgrvg6z4o.amp.

[17] https://www.theguardian.com/us-news/2025/oct/29/times-de-blasio-fake-interview-mamdani.

[18] https://www.theguardian.com/politics/2025/oct/18/tory-mp-reports-ai-generated-deepfake-video-of-him-announcing-defection-to-reform-uk.

[19] https://www.bbc.co.uk/news/live/cd9kdl54x9lt.

[20] "When Cracker Barrel changed its logo in August 2025, 44.5% of initial social 'media outrage' was bot-generated." https://finance.yahoo.com/news/cracker-barrel-logo-controversy-driven-161940333.html?guccounter=1&guce_referrer=aHR0cHM6Ly93d3cuZ29vZ2xlLmNvbS8&guce_referrer_sig=AQAAADrm4JQB0E6L3G19CDA00Qr1JRBfX0mfVr0ZfYo_dvXhfMYWUjiYN0AMeyxJ-1Q7OqQJg3BoveJNhrtvuj1FuIG_dmkmFs01yLiYKMTwu2iJf2eAiIki3AiLJAGn_o1rlxbzVW61-Zyh1WARd8i9oNkrLdshXoFcMaqSkPa4xHyx.

In an astonishing prediction, World's blog estimates that by 2026, '90% of the content on the internet will be AI generated.'

Deepfakes are becoming harder to detect and easier to access.[21] In the hands of the citizen, the potential impact of Deepfake-enabled surveillance tools on policing is a cause for concern; in the hands of a malign state, the capability to generate objectively incontrovertible biometric evidence of wrongdoing by its citizens is a terrifying prospect not even foreseen by Orwell & Co.

With AI 'inundating the internet, crippling our political structures, and undermining reality'[22] and most of it visual, the traditional veracity of the camera has long been extinct.

Clearly, the rise in AI-enabled fraud presents policing with a huge challenge. Deepfakes are being used non-consensually in many frauds and scams, with generative AI reproducing biometrics like voices.[23] National law enforcement organisations like Europol and the US Department for Homeland Security[24] have published substantial analyses and guidance on the threats from Deepfakes. The weaponisation of Deepfakes in terrorism[25] has led to 'deepfake literacy' programmes being delivered as part of wider countering violent extremism strategies which are even considering the prospect of using deepfakes to undermine the conditions that allow extremist groups to thrive (Busch & Ware, 2023).

So, what can be done in response to the bare-faced lying of the Deepfake and the risk to AI-enabled surveillance?

The first thing might be to acknowledge that we are not dealing with cameras anymore and haven't been for some time. Deepfakes are not photography; their images aren't photographs and the old 'fairies at the bottom of the garden' camera hoaxes that seem so naive now[26] are from a bygone age.

---

[21] https://me-en.kaspersky.com/about/press-releases/kaspersky-warns-deepfake-services-are-now-400-times-cheaper-and-more-accessible.

[22] https://futureoflife.org/project/combatting-deepfakes/.

[23] Future of Life ibid.

[24] https://www.europol.europa.eu/cms/sites/default/files/documents/Europol_Innovation_Lab_Facing_Reality_Law_Enforcement_And_The_Challenge_Of_Deepfakes.pdf; https://www.dhs.gov/sites/default/files/publications/increasing_threats_of_deepfake_identities_0.pdf.

[25] https://icct.nl/sites/default/files/2023-12/The%20Weaponisation%20of%20Deepfakes.pdf.

[26] Faked photos of the Cottingley Fairies even had the creator of the world's greatest detective fooled. https://www.bbc.co.uk/news/uk-england-leeds-55187973.

The next thing is to recognise the adaptability of Deepfakery. Deepfakes are omni-purpose and can be used to fuel every conceivable human motivator—antagonism, materialism, deism, eroticism, dogmatism (see e.g. Rahman, 2024; Toczyki & Banerski, 2021). It seems that there is no theory too absurd nor ideology too repellent to find some corroboration somewhere in the online world[27] and Deepfakes can feed the confirmation biases of the credulous.[28]

Further, we will have to accept that the only way to detect Deepfakes is with more technology. Public awareness is growing and the response from the tech sector and governments is producing fact-checking sites and open competitions trying to keep up. But in the end, we're going to need assistance from AI. We should enter this phase with caution—Deepfakes and their detection are rapidly becoming big business[29] and we should be sceptical if co-labs creating new poisons try to sell us the antidote.

Perhaps the most important thing we can do at this stage is to prepare for what's coming. Believing the output from a smart device as unadulterated truth is technologically naïve; the camera is long gone and has taken its reputation for candour with it. In the third era of surveillance, however, the adage may make a comeback.

However endlessly adaptable its progeny may have been, the device of the second era was a tool in the hands of an active operator. We're still the ones doing the lying. In the next era the surveillance device will morph from interactive instrument to autonomous creator. Without need of a human operator, the agentic AI surveillance entity will be capable of deception of its own volition. In that scenario the 'camera' *will* be able to lie to us and, in a state surveillance context, that is something we haven't begun to grapple with.

References

Akhgar, B., Staniforth, A., & Bosco, F. (Eds.) (2014). Cyberspace: The new frontier for policing? In *Cyber crime and cyber terrorism investigators' handbook* (Chapter 1). Elsevier.

Busch, E., Ware, J. (2023). *The weaponisation of deepfakes—Digital deception by the far right*. International Centre for Counter-Terrorism.

[27] https://www.bbc.co.uk/news/blogs-trending-38156985; https://www.livescience.com/11375-top-ten-conspiracy-theories.html and targeted.

[28] https://www.scientificamerican.com/blog/observations/why-are-deepfakes-so-effective/.

[29] https://www.biometricupdate.com/wp-content/uploads/2025/10/BU-Deepfake-Detection-Market-Report-and-Buyers-Guide.pdf.

Rahman, T. (2024, September 8). How online interaction foments odd or dangerous beliefs. *Psychology Today.*

Toczyki, P., & Banerski, G. (2021). Patterns of digital users' irrationality in the moment of digital transformation of the media. In C. Bieli, J. Kacprzyk, J. W. Owsiński, R. Romanowski, & M. Sikorski (Eds.), *Digital interaction and machine intelligence* (pp. 115–124). Springer.

15 Whatever Happened to Privacy?

What does privacy mean to you? Is it a right to keep some things hidden or secret, or more of an ability to do so? Perhaps it's a freedom from intrusion or inspection. Maybe it's about respecting boundaries. In the context of surveillance privacy is often about being seen. Consider the following scenario:

> On 20 August 1995 at 11.30 p.m., a man suffering from depression as a result of personal and family circumstances walked alone down the high street towards a central junction in the centre of Brentwood, England, with a kitchen knife in his hand, and he attempted to commit suicide by cutting his wrists. He stopped at the junction and leaned over a railing facing the traffic with the knife in his hand. He was unaware that a CCTV camera, mounted on the traffic island in front of the junction, filmed his movements. The CCTV footage later disclosed did not show the man cutting his wrists, the operator being solely alerted to an individual in possession of a knife at the junction.
>
> The police were notified by the CCTV operator and arrived. They took the knife from the man, gave him medical assistance and brought him to the police station. He was detained under the Mental Health Act 1983. His custody record refers to the self-inflicted injuries to his wrists on arrival and notes that he was examined and treated by a doctor, after which he was released without charge and taken home by police officers.

Has this man's privacy been respected? Do we have any expectation of privacy when we walk down a street? We can be seen by others who are also in or overlooking public places, but does that mean we can be filmed and recorded? Should our expectations change if we're carrying a knife or are in need of help?

Let's see what happens next.

F. Sampson, *How AI-Driven Surveillance is Changing our World*, Security Informatics and Law Enforcement, https://doi.org/10.1007/978-3-032-24005-7_15

> On 14 September 1995 the CCTV working party of the Council agreed to authorise the release of regular press features on the CCTV system. The Council also decided to cooperate with third parties in the preparation of factual programmes concerning their CCTV system. The Council's first press feature, the *CCTV News*, was released on 9 October 1995 and included two photographs taken from the CCTV footage of the man to accompany an article entitled 'Defused—The partnership between CCTV and the police prevents a potentially dangerous situation'. The man's face was not specifically masked. The article included the name of a Council employee in the event that readers wished to obtain copies of the pictures.
>
> On 12 October 1995 the *Brentwood Weekly News* used a photograph of the incident involving the man on its front page to accompany an article on the use and benefits of the CCTV system. The man's face was not specifically masked.

Has the man's privacy been breached now? If so, in what ways and by whom? Did they have an excuse or justification? Pictures and news stories like this appear all the time, and the council and the police will often want to promote the 'effectiveness' of their investment in surveillance technology.

The story continues:

> On 13 October 1995 an article entitled 'Gotcha' appeared in the *Yellow Advertiser*, a local newspaper with a circulation of approximately 24,000. The article was accompanied by a photograph of the man taken from the CCTV footage. The newspaper article referred to the man having been intercepted with a knife and a potentially dangerous situation being defused as a result of the CCTV system. It was noted that the man had been released without charge.
>
> As a result Anglia Television sought, and the Council provided, footage of the incident involving the man. On 17 October 1995 extracts from that footage were included in its news programme about the CCTV system, a local broadcast to an average audience of 350,000. The man's face had been masked at the Council's oral request. However, that masking was later considered inadequate by the Independent Television Commission, the man's distinctive hairstyle and moustache making him easily recognisable to anyone who knew him.
>
> On 18 October 1995 the Chairman of the Council informed the Council Technical Services Committee that cooperation had been, and would continue to be, given in the preparation of factual documentary programmes concerning the CCTV system. He referred to the feature on CCTV which had been broadcast by Anglia Television on the previous day.
>
> In late October or November 1995 the man became aware that he had been filmed on CCTV and that footage had been released because a neighbour told his partner that the former had seen him on television. He did not take any action then as he was still suffering from severe depression.

> On 16 February 1996 a second article entitled 'Eyes in the sky triumph' was published in the *Yellow Advertiser* outlining the benefits of CCTV in the fight against crime and was accompanied by the same photograph as had been previously used by that newspaper.

By this point you might start to think that the man's privacy has not just been intruded upon but maybe trampled. When he tried to challenge the council's use of surveillance images in the High Court, he failed. Relying on his right to respect for a private and family life,[1] he subsequently brought his case against the UK before the European Court of Human Rights in Strasbourg.[2] The Court held that the disclosure of his images by the council was a breach of his entitlement to respect for his private life. The Court restated a previous judgement in a different case against the UK,[3] namely that there is a 'zone of interaction of a person with others, even in a public context, which may fall within the scope of *private life*.' The local authority's disclosures of the CCTV images had made a moment of intensely personal behaviour at a time when he was suffering from severe mental impairment viewable to an extent which far exceeded any exposure to a passer-by or even to CCTV security personnel. The images were viewable to a degree that surpassed anything the man could possibly have foreseen when in the street and the disclosures were a serious and unjustified interference with the man's right to respect for his private life.

People (often lawyers, politicians and journalists) talk loudly about the need to protect privacy from surveillance, but before you can protect something or regulate activity that threatens it, you need to be able to define it—and there are several problems with identifying the parameters of privacy. Even before the arrival of AI-driven surveillance tools, social media saturation of our lives and the infinite story cycling of 'news', privacy has been a particularly slippery legal concept (see Solove, 2008). Courts have struggled to define exactly what privacy looks and feels like, and a lot of time and money has been expended fighting over it. Answering questions like 'what is a reasonable expectation?', 'how does this apply to the already famous' or 'what about vulnerable people who are in need of protection?', the law has provided some help in understanding where the limits are, but it hasn't kept up with surveillance technology or our insatiable use of it.

---

[1] Under article 8 of the European Convention on Human Rights and Fundamental Freedoms.

[2] *Peck v United Kingdom* (2003) 36 EHRR 41—the wording in the scenario is taken directly from the published judgment of the Court.

[3] *PG and JH v United Kingdom* (2008) 46 EHRR 51).

If, like Mr Peck in the above scenario, your privacy is 'breached' by public surveillance, what can you do about it? Bringing the case against the government means your name (and therefore the incident you wished to keep private) is part of the public record. One cost of using the law to pursue your right to privacy is your loss of anonymity.[4] The right to an effective remedy is also an entitlement under the European Convention,[5] but what does that mean in the context of state surveillance? Is financial compensation enough[6] or should you be able to change to require punishment for, or a change in behaviour of, the police and local authorities?

In terms of the law, citizens in some countries enjoy direct protection of their privacy by rights enshrined within their constitution, countries such as the United States of America.[7] Some—like Australia and New Zealand—have senior officials appointed as privacy commissioners, acting independently of the state with responsibilities for reviewing surveillance activities, publishing formal guidance and punishing breaches. Citizens in the UK have no free-standing right to privacy[8] and no privacy commissioners. Instead they enjoy an extension of the wider legal concept of confidentiality,[9] adapted to incorporate Article 8 of the European Convention on Human Rights protecting entitlements to the enjoyment of home and family life etc. and, if the domestic law won't protect them, they can bring an action against the country as Mr Peck did—although it's important to look at the dates in his case: breach by local authority 1995—decision by the ECtHR 2003.

Debates about surveillance usually centre around some notion of privacy, but many take place without ever establishing what people believe their privacy is or should be. In the AI-enabled future, we will need to be clearer. We have seen how the traditional concept of CCTV is no longer typical of public space surveillance,[10] and how the extent of intrusion of AI-enabled surveillance turbocharges the 'viewability' of images (including live video streaming) in comparison to the already unjustified extent in Mr Peck's case. AI-enabled technology like retrospective facial recognition means that any images of Mr Peck in existence before or since the event—school photos, stills of crowds at sports fixtures and public

[4] With notable exceptions around age and sexual offence cases.

[5] Article 13.

[6] Mr Peck was awarded €15,075, worth around €24,000 today.

[7] See for example the Fourth and Fourteenth Amendments.

[8] *Campbell v MGN* [2004] 457.

[9] Along with a tortious cause of action for some infringements.

[10] Chapter 5.

gatherings, transport networks—can be searched for, found and matched cheaply and easily, while internet scraping will have a very good chance of finding where he is today. If he is no longer alive, any legal rights protecting Mr Peck's personal data and reputation will have died with him: UK privacy and data protection laws only apply to the living, even when the police or criminals intrude on the dignity of the deceased in the most appalling circumstances.[11]

Our own private use of sophisticated technology has also 'socialised' surveillance (see e.g. Lyon, 2017) completely changing both the notion of privacy and any reasonable expectations of it.

In settings like home security, international travel and online purchasing, we can see a chicken-and-egg causal loop, where consumer convenience is driving remote surveillance-focused innovation and vice versa.[12] Advanced surveillance technology offers us many convenient ways of keeping an eye on things from afar: houses and property, vulnerable relatives, gardens, etc. And the convenience is not all for the consumers'—the mining of big data made available by online behaviour and digital engagement provides an extraordinarily potent surveillance capability, as illustrated by other cases.[13]

If they were difficult to draw before AI-powered surveillance capability, the boundaries around privacy will be even more elusive in the future. Technological and societal changes have given the state many highly effective ways of keeping tabs on us, and we don't need to be 'visible' or 'present' in a public place for our information revealing our and past behaviour to be known and knowable. By far the biggest source of this surveillance data is not state-operated CCTV apparatus, but our own devices. Technology consumerism has transformed state surveillance way beyond the government's procurement of surveillance equipment. A perfect example is the phone in your pocket, on your desk or by your bed.

It is almost inconceivable that an economically or socially engaged person will manage for long without a mobile phone. Whenever and wherever they do so,

[11] Two former Metropolitan Police Service (MPS) officers have been jailed following a hearing at the Old Bailey today (6 December) for taking and sharing photographs from the scene of Nicole Smallman and Bibaa Henry's murders. https://www.policeconduct.gov.uk/news/former-metropolitan-police-service-officers-jailed-taking-photographs-crime-scene; Double murderer David Fuller abused the bodies of at least 100 women and girls in two hospital morgues over a 12-year period and kept videos of his actions. https://www.bbc.co.uk/news/videos/cx2nx7dzrrvo.

[12] Chapter 7.

[13] Like *Lloyd v Google LLC* [2021] UKSC 50 where the search engine secretly tracked the internet activity of millions of iPhone users and used the data for commercial purposes without the users' knowledge or consent.

the person will be susceptible to tracking by cell site location information (CSLI) and global positioning satellite (GPS) monitoring. It may not offer images but the surveillance data gathered by the state through CSLI and GPS is, in the powerful description by the US Supreme Court, 'detailed, encyclopaedic, and effortlessly compiled'.[14] In that setting, the Supreme Court decided that privacy meant where a citizen 'seeks to preserve something as private'. Where the police use CSLI and GPS to track someone's past movements, the US Constitution regards that as state intrusion into the person's privacy and as such it must be supported by official authority (in Mr Carpenter's case, a warrant).

In *Carpenter*, the government obtained several months' worth of phone location records for people suspected of crime in the course of an investigation in Detroit. They did not get a 'probable cause warrant' first. Mr Carpenter's records contained almost 13,000 separate data points of his phone's location, averaging more than 101 per day over a period of four months, showing his location at the time, including when and where he attended places of worship. State monitoring of individual activity to this level engages fundamental rights and freedoms such as freedom of religion, assembly and association, and the Court in *Carpenter* recognised that such information must be protected by the Fourth Amendment.

The law and principles covering direct surveillance by the state are relatively mature in many countries and there are clear statutory requirements for obtaining prior authority,[15] but these usually apply only where the intrusive behaviour—including compulsion of records—has come from the police. It's not just our own phones that are giving our movements away—our fellow citizens and their devices are also rapidly becoming a source of surveillance data. Sharing the output of our smart doorbells and dashcams and Internet of Things (IoT) connected appliances[16] with the police gives them a formidable—and often undetectable—remote surveillance capability. Photo location apps capable of identifying where a picture was taken using nothing more than the image and facial matching software were beyond the reach even of state intelligence agencies when Mr Peck was being filmed in the street. A privacy consequence of this trend is that state surveillance is likely to be more intrusive, less conspicuous and harder to regulate. Self-surveillance is more intrusive than anything the police try to operate directly because the lines drawn around the processing of our personal data are much more restrictive for the police and law enforcement than those constraining what we do as individuals. From a policing perspective, if citizens choose to

[14] *Carpenter v United States* 2018 No. 16–402—see also *Katz v. United States* 389 U. S. 347.

[15] The Regulation of Investigatory Powers Act 2000.

[16] Chapter 2.

make the content of their private devices available, why wouldn't the police use it for proper policing purposes? How could they decide which images to ignore?

The American Civil Liberties Union (ACLU) described *Carpenter* as one of the most consequential rulings regarding privacy in the digital age.[17] In that case the SC added that a protected expectation of privacy is 'one that society is prepared to recognise as reasonable'. This is interesting, given the extent of our digital profligacy as demonstrated by widespread use of social media and our online behaviour, which can have a paradoxical effect on privacy[18] What all this means for the 'reasonable expectation of privacy recognised by society' is far from clear, but unless the oversight and accountability mechanisms governing police surveillance are extended to include these capabilities, their effect will be diluted and regulation will be increasingly difficult to achieve.

Meanwhile, the value from surveillance for the police comes in the form of its *product*. Counterintuitively, our giving them access to our aggregated surveillance capability may mean the police may need fewer cameras in the future.

How this symbiotic surveillance ecosystem will evolve is uncertain, but it has already presented the digitally exposed citizen with a very different world from that of the CCTV camera where we began this chapter, and on which most of our regulatory frameworks are still based.

In the future, will the 'old-world legal rules' automatically apply to new forms of digital surveillance? Not according to the US Supreme Court.[19] However, where the police trial new surveillance technology like live facial recognition, any interference with the citizen's Article 8 rights will still need to be in accordance with the law.[20] And it will be important for their AI accountability more broadly that the police are able to demonstrate the lawfulness of their use.

The UK has seen some significant legal challenges to the use of state surveillance where it has interfered with the citizen's legitimate expectations of privacy. The use of new police surveillance technology (body worn video) by officers

---

[17] https://www.aclu.org/news/privacy-technology/supreme-courts-most-consequential-ruling-privacy-digital.

[18] As can be seen by the creation of the 'Streisand effect' whereby attempting to suppress a disclosed item simply draws more attention to it: What Is the Streisand Effect? (The Economist, Apr 15, 2013).

[19] *Carpenter loc. cit.*

[20] Either an interference is in accordance with the law or it is not—*R (on the application of Bridges) v Chief Constable of South Wales Police* [2020] EWCA Civ 1058; the issue of whether there is relevant 'law' for this purpose is a binary question: see *In re Gallagher* [2019] UKSC 3.

when they are present in someone's home is one example[21]; the intrusive surveillance of journalists in Northern Ireland another.[22] These cases have advanced the body of law governing state intrusion and privacy, but, at the same time, have damaged public confidence (Higgins & Halkon, 2023).

What's the *opposite* of privacy? Total visibility? Being under constant surveillance? The US Supreme Court in *Carpenter* held that government access to detailed location data from CSLI and GPS provided a method of 'near-perfect surveillance'; in the few years since, it has got a lot nearer. State access to surveillance data from satellites, IoT devices and AI-enabled capabilities has brought us ever closer to 'omniveillance'.[23] Darwinian scholars know that scarcity is the precursor to extinction and any traditional notions of privacy are vanishing fast. Our traditional wish for privacy is rapidly losing out to our need to be noticed (Sampson, 2026) and the privacy question for the AI-enabled future will be how far *avoiding* being seen is even possible, let alone expected.[24]

References

Higgins, A., & Halkon, R., (2023). *Contact and confidence in a digital age: Improving police-public relations with technology.* Police Foundation.

Lyon, D. (2017). Surveillance Culture: Engagement, Exposure, and Ethics in Digital Modernity. *International Journal of Communication*, 11, 824-842

Sampson, F. (2025). https://www.biometricupdate.com/202510/transplanting-biometrics-in-police-cctv-may-create-frankenstein-effect

Sampson, F., (2026) 'The future of police surveillance and cameras', Chapter 16 in Future Policing, Technology and Transformation Huey, L., Ferguson, L. and Adams, I. (Eds), (2026) Routledge, New York

Solove, D. J. (2008). *Understanding privacy.* 1Harvard.

[21] *AP v Hampshire Constabulary* UKIPTrib IPT 17 191 C 5 February 2019.

[22] *Lee and Wilkes v Security Service* IPT/22/12/H and IPT /22/11/H, 17 December 2024.

[23] Chapter 4.

[24] Explored in Chap. 16 Do we have a right to go missing?.

Do We Have a Right to Go Missing (and Would It Be Worth Having If We Did)?

16

Do we have a right to go missing?

In surveillance terms this question is twinned with the privacy question[1] and we know how advances in AI-enabled technology have created a global explosion in surveillance capability, spurring concerns about intrusion and anonymity. The combination of remote biometric technology, our dependence on inter-connected digital devices and the ability to combine, sort and match the data from these layers has given the police extraordinary surveillance power. It's also given them a dilemma—who should they look for?

In the law enforcement world there are plenty of people who don't want to be 'found', principally people who are 'wanted' and people who have gone missing under circumstances that cause concern for their wellbeing. In both cases the police have a duty to look for them.

Near total digital visibility presents a big problem when the state needs, not just to respect anonymity, but to *guarantee* it. For some people, not being found is not a privilege or a whim; it's critical to their safety and that of others and it's increasingly elusive. For example, complainants rebuilding their lives after abusive relationships need a significant degree of anonymity if they are to have the space and confidence to recover. Witnesses in serious organised crime cases may need ongoing protection by the state, as will some prisoners released after serving sentences for notorious offences. All of these individuals within the justice system face significant risk if 'found', but it is becoming harder to guarantee that they won't be. Undercover law enforcement operatives also fall into this group. Providing new identities was a tactic in the past, but this has

[1] Chapter 15.

F. Sampson, *How AI-Driven Surveillance is Changing our World*, Security Informatics and Law Enforcement, https://doi.org/10.1007/978-3-032-24005-7_16

become much more challenging with the introduction of biometric proof of ID requirements generally, while the accountability of officers allowed to operate behind false identities has raised serious concerns.[2]

What about the rest of us? Do we have a general right to go missing if we choose to?

The thing with rights is that there's usually something standing in the way. Even fundamental rights like the entitlement to freedom of speech or association are usually 'qualified' in some way. Someone else may enjoy an equally fundamental right that's in conflict or competition with yours—and the state may have a legal duty that necessarily places limitations on both. Policing provides many examples of this, as we've seen above, but what about those who just want self-determination, people impelled by mid-life crisis or New Year resolution to grab a bag and head off over the horizon—do they have a right to go off grid?

If today you decide to do that in the UK, you don't have any specific 'right' that you can point to. On the other hand, the police have no general obligation to go and look for you. So far, so good. But even though there is no duty for the police to look for, still less to reveal the location of someone who has chosen to go 'no contact', how can they be sure your absence is voluntary and benign? What started out as an exercise of liberty to go missing after, say, you finished writing a book about surveillance may have led to you having an accident, falling ill, becoming a victim (or perpetrator) of a crime or some other subsequent jeopardy unforeseen at the outset of your bid for freedom. Your disappearance may have been caused by a scam or coercion. All those 'don't worry about me' messages you sent to loved ones, voice notes and social media posts letting everyone know you were seeking solitude, turn out to be deepfakes. Sometimes people send messages before going dark because of something murky they have done or are planning to do. In all the above situations, the police have a duty at least to look more closely, and policing experience shows it's the one you decide not to look for that will be the one you regret. Even where they *do* look for missing people, the police can attract significant criticism and distraction.[3] For the police, the looking is the easy bit—all the risk lies in not doing it.

Your chances of being able to disappear without further state interest will certainly be hindered if you have any 'indicators of vulnerability' (particularly age and mental health). In an age where we are permanently connected to communication channels and social media visibility, lack of contact is itself likely to be a trigger for further investigation and is often seen as suspicious or abnormal.

---

[2] https://www.ucpi.org.uk/.

[3] https://www.college.police.uk/article/review-nicola-bulley-investigation-published.

Paradoxically, asserting your right to go missing has become proof that you need to be found.

Perhaps this is more of an extension of privacy than a free-standing entitlement—and we know privacy is almost indefinable let alone attainable.[4] If there's a freedom here, however heavily circumscribed, we need to understand what it is we are trying to be free from (Russell, 1922). In asserting an entitlement to go missing, what is it that we're trying to be free of? If it's a negative concept—not being looked for—that will bring the above police duty into play but it will also attract further restrictions. First, it is not only the police that can feel an obligation to look for you. There can be as many reasons to go looking for someone as there are to go missing: remorse, reparation, restitution; intestacy, genealogy, philanthropy; confession, absolution and closure. This is why there are charities dedicated both to finding people who have gone missing and also helping people to go missing without being found, in each case minimising alarm and the unnecessary expenditure of resources. The prevalence of absence has resulted in commercial companies creating huge databases of 'missing persons', ready to join in the activity. We may be at liberty to go missing but we probably won't be free from a lot of well-meaning searching when we do.

Secondly, the grid doesn't have many off-ramps any more. By severely limiting how far you'll get before triggering something, the technology we use in our everyday lives has made finding you easier than not finding you. Police crime scene investigators live by an old forensic science acronym ECLAT—every contact leaves a trace—and it is as applicable in the setting of mass surveillance as anywhere. Each and every digital interaction we make will create a digital spoor, leaving something findable by someone. If you take your phone, even a 'burner' will generate detailed location data from cell site location information (CSLI) and global positioning satellites (GPS). These data points provide the state with a method of surveillance that the US Supreme Court described as 'near-perfect'[5] and the information they contain is 'detailed, encyclopaedic, and effortlessly compiled'. Other sophisticated telemetry that we've adopted in our trainers and fitness watches and navigational aids will also blow the digital whistle on our whereabouts.

If you plan on using a vehicle in your bid for freedom, there is better news. The UK's automated number plate recognition (ANPR) network covers most of the country's roads and collects tens of millions of reads per day, but here your ability to avoid the surveillance technology is very high. Like most countries that

---

[4] Chapter 15.

[5] *Carpenter v United States* 2018 No. 16–402—see also *Katz v. United States* 389 U.S. 347.

rely on vehicle-borne credentials for identification and surveillance purposes, the UK system is extraordinarily vulnerable (Sampson, 2022). The plate itself is a single point of failure in the system and cloning them—or, even simpler, applying cling wrap or making some other minor change to the face of the plate—renders plates illegible by the cameras.[6]

Notwithstanding the shortcomings of any national vehicle identification schemes, accessing many of the bare essentials you will need will expose you digitally. Filling up your unidentified car will bring you into contact with security systems at forecourts even if you pay with cash. Once you exhaust your cash reserves, ATM withdrawals from accounts registered to your name will trigger instant alerts, while accessing any online services, seeking help and assistance, and communicating with others while roaming freely will generate CCTV and surveillance data that will be recorded and recoverable forever. Requirements for proof of ID pretty much cover any contractual services and if your break for the hills includes travel beyond your country's borders there will be an elaborate system of measures (including advanced biometrics) for recording the fact that you left, when and where you were going next. And there will be a correlative set of rules waiting at your destination. Realistically, any 'freedom' we might enjoy from the need to register our movement with the state only exists in any practical sense if we stay within the country where we currently find ourselves.

Thirdly, buses, trains, railway and coach stations are pretty well covered in terms of sited surveillance cameras, but this isn't how the police find you anymore. Public space surveillance is no longer about where the police put their cameras; it's about what they do with the images from everyone's cameras. Our global obsession with uploading images of our every moment and movement has redefined the concept and boundaries of 'mass surveillance' and the state can tap into that citizen capability. Finding a needle in a haystack has changed from a simile of the impossible to simplicity itself—AI-enabled surveillance excels at finding specific needles in vast digital haystacks, and our ability to do it comes less from the state having cameras and more from the fact that we're all looking at everything and sharing what we see.

The precise power of this aggregated capability has yet to be understood, but there's no doubting its impact. In January 2025, police in England launched a fresh appeal to find a woman they had been looking for since 1972, when she went missing at 16 years old. According to the Guardian,[7] the police uploaded a

---

[6] For a full discussion of how and why see Chap. 13.

[7] https://www.theguardian.com/uk-news/2025/jan/01/coventry-woman-who-vanished-52-years-ago-found-alive-and-well.

photograph on social media, prompting a very quick public response. The police were able to confirm that she was alive and well and living in another (undisclosed) part of the country, that her disappearance had been voluntary and that no harm had befallen her. The aggregated surveillance power of the UK had empowered the police to locate in a matter of hours someone who had enjoyed freedom from state surveillance machinery for the past 52 years.

When we add open-source satellite data to the picture, the ability of the state to shrink our worlds in the pursuit of persons of interest is so potent, it would make Robert Clayton Dean[8] shudder. The power of open-source surveillance is dramatically illustrated by several long manhunts for mafia figures who had been able to live below the police for decades. In one such case, an Italian mafia boss who had escaped from prison in Rome in 2002, where he was serving a life sentence, was captured in 2022 after detectives spotted him on Google Street View[9] In another, a mafia fugitive was caught after appearing in a cooking programme on YouTube in 2021, having evaded detection for the previous seven years.[10] If organised crime bosses are unable to avoid the eye of the state, the chances of the average citizen doing so are not very high.

As human traceability becomes endemic, other rights assume a greater importance. Data rights, protections from covert and directed surveillance, international sharing and storage of images and biometrics are all essential for mitigating intrusion. Encryption of communications can offer a shield against content interception, but ultimately the state has to balance freedoms against the threat of wider harm, as some phone companies are discovering,[11] and sometimes guarantees of confidentiality can be overridden by the public interest in, for example, investigating murders or terrorism (Sampson, 2016).

One thing that I am 'free from' is any general obligation to reveal my face or identity, and you could avoid many facial recognition surveillance tools by covering up your face. The police in the UK have specific powers to require you to remove a face covering but they are exceptional and limited to narrow circumstances.[12] Despite the concern at the government's proposals to make it a criminal offence to conceal your face in public, the power will only apply in

[8] The principal character in Enemy of the State played by Will Smith who is relentlessly pursued by intelligence agencies https://www.imdb.com/title/tt0120660/.

[9] https://www.bbc.co.uk/news/world-europe-59884803.

[10] https://www.theguardian.com/world/2022/jan/05/italian-mafia-fugitive-arrested-in-spain-after-google-maps-sighting. This surveillance capability is one reason why it is impossible to calculate the 'world's most surveilled cities' and other lists—as explained in Chap. 4.

[11] https://www.bbc.co.uk/news/articles/c798xv5qwylo.

[12] See the Criminal Justice and Public Order Act 1996, s.60AA.

small geographic areas temporarily designated for public protest where the police have given the public notice of it being used.[13] Unless you wander into an area where the police have given such notification, you're probably safe keeping on your mask and baseball cap when on the streets of the UK. Nevertheless, finding people has dominated police investment in surveillance technology over the past decade. When the College of Policing for England and Wales advises that the use of biometrics like FRT can be justified if the police are looking for 'potential witnesses' (a wholly indeterminate term which includes everybody), any freedom from being sought is under increasing threat.

Do we have a right—or freedom—to go missing? Possibly, but if you're thinking of setting off on a voyage of self-discovery without a trace, that ship has already sailed, taking its passenger manifest with it. If you do feel the urge to head out into the world to 'find yourself', someone will almost certainly have beaten you to it.

Your journey through your home country may take you past local authority CCTV cameras unnoticed, but some are now being retrofitted with biometric capabilities.[14] In addition, how many houses and offices with smart doorbells will pick you up? How many drones? How many times have you unwittingly starred in a walk-on part in the background of selfies, dashcams of passing vehicles and bikes with GoPros? The answer is that we won't know, but the ability of AI-enabled technology to sort through all images captured in this way increases, not just since you began this chapter but since you began this sentence. Expectations of privacy are difficult to define and far from unlimited, but our digital visibility is almost absolute. In a world where we're so easily found, a right not to be is like a right to modesty where all clothes are see-through: entirely dependent on no one looking.

References

Russell, B. (1922). *Free thought and official propaganda*. Huebsch Inc.

Sampson, F. (2016). Whatever you say… ': The case of the Boston College tapes and how confidentiality agreements cannot put relevant data beyond the reach of criminal investigation. *Policing: A Journal of Policy and Practice, 10*(3), 222–231.

[13] Crime and Policing Bill—https://www.gov.uk/government/publications/crime-and-policing-bill-2025-factsheets/crime-and-policing-bill-public-order-factsheet.

[14] Creating the risk of a Frankenstein effect—https://www.biometricupdate.com/202510/transplanting-biometrics-in-police-cctv-may-create-frankenstein-effect.

Sampson, F. (2022). *Annual report of the Surveillance Camera Commissioner 2021–22 House of Commons Library*

What We Talk About When We Talk About Police Biometrics

17

When I was the UK's Biometrics and Surveillance Camera Commissioner, I wrote a blog focusing on what we talk about when we talk about biometrics. That was in 2022 and the UK was talking a lot about police biometrics. In particular it was talking about diluting an already thin regulatory framework for biometrics and surveillance, while the police were experimenting with new AI-enabled technology like facial matching and Live Facial Recognition. It was a time when the courts were also talking about police biometrics and the need for clear legal frameworks and the risks of retention of images and other sensitive personal data.

The blog was drawn from my formal response to the government's statutory consultation on scrapping the need for it to publish a code of practice regulating public space surveillance by the police and local authorities.[1] The legislative plans fell by the wayside when the prime minister called a snap general election in 2024 and the proposed regulatory framework for police biometrics stalled. Following a series of interim measures, the government appointed a new commissioner on 1 November 2025.[2]

Since that time the technology has continued to evolve relentlessly, and public expectation of its use by the police in the UK has moved with it,[3] but where

[1] https://www.gov.uk/government/publications/data-a-new-direction-commissioners-response/dcms-consultation-data-a-new-direction-response-by-the-biometrics-and-surveillance-camera-commissioner-accessible-version.

[2] Professor William Webster—https://www.gov.uk/government/news/new-biometrics-and-surveillance-camera-commissioner-announced.

[3] https://aipas.co.uk/.

F. Sampson, *How AI-Driven Surveillance is Changing our World*, Security Informatics and Law Enforcement, https://doi.org/10.1007/978-3-032-24005-7_17

has that left regulation? Reports suggest[4] that the pace of policy and reform in biometrics internationally continues to be frustratingly out of step with both the technology and our expectations of its use. The gap between the possible (what can be done), the permissible (what must/must not be done) and the acceptable (what we support being done) has widened, while the opportunity to close it is narrowing. With renewed admiration for Raymond Carver,[5] it is therefore timely to listen to what we talk about now when we talk about police biometrics—and what we should probably be talking about next.

When people talk about biometrics and policing, many still talk about the advent of a dystopian, made-up world, one where state-run surveillance is the great terror and "Orwellian" the adjective of choice. But that's not where police biometric technology is taking us[6] and the lines of separation between state-operated surveillance machinery and our everyday lives are rapidly receding.

When we, the policed, talk about police biometrics, we should really be talking about the fact that the world has turned 90°, moving from its traditional citizen-facing aspect where the police collected images *of* the people to one where they collect images *from* the people. We should be talking about how we now have access to biometric surveillance technology that only a few years ago was the preserve of state intelligence agencies—we're talking about AI-powered spectacles with always-on face recognition and lip-reading functionality, about liveness detection apps and Wi-Fi routers that can see people through walls. We should be talking about how the traditional surveillance face-off between privacy and security is being replaced by convenience and public interest.[7] What does that mean for our constitutional freedoms[8] for police accountability[9] and for the future?[10]

When we talk about police biometrics, we talk about the intrusiveness of the technology but not so much about the intrusiveness of damaging behaviour that biometrics are already preventing. We talk less about how, once the police adopted biometrics like DNA and fingerprints, the techniques revolutionised criminal investigation and how a past where they weren't available to policing is

[4] https://www.biometricupdate.com/202509/biometrics-inspire-trust-policy-makers-invite-backlash.

[5] The title of the blog and the chapter are reflections of Carver's short story What We Talk About When we Talk About Love—https://en.wikipedia.org/wiki/What_We_Talk_About_When_We_Talk_About_Love.

[6] As the other chapters of this book illustrate.

[7] Chapter 7.

[8] Chapter 15.

[9] Chapter 10.

[10] Chapter 20.

difficult to imagine, still less remember. We talk hardly at all about how roadside substance testing and back-calculation techniques have made our roads significantly safer.

When we talk about police biometrics, we may not talk about regulatory frameworks very often (or often enough), but when we do, we're probably thinking about something designed for a different age. We talk about 'guard rails'[11] and 'standards'[12] based on the risks of a pre-AI era when public spaces were watched over by visible, publicly operated cameras looking down on us from fixed points. We don't yet talk about how to safeguard our integrated biometric surveillance data stored on mega-databases and processed by data centres that will consume more electricity than the cities where they're sited,[13] or how to protect them from cyber-attack and insider threats.[14] We don't yet talk about the next steps in our growing data sharing with the state, about moving from giving up the 'product' of all our citizen-controlled biometric machinery and being directed by the police to livestream and capture images at the scene of an incident before they arrive.[15]

We should talk about how even the EU AI Act[16] misses this evolutionary phase and why using 75-year-old fiction to predict the future of AI is an ironic failure of imagination.

When we talk about police biometrics, we often find ourselves talking about how inaccurate facial recognition algorithms used to be rather than the almost 100% accuracy rates now being achieved,[17] and the fact that people generally trust the technology when it's in their own hands. When we talk about whether we trust the police to use biometric surveillance responsibly, we should recognise that this isn't a technology question; it's an accountability question.

---

[11] https://www.nytimes.com/interactive/2025/09/15/opinion/nypd-surveillance.html.

[12] https://post.parliament.uk/use-of-digital-scientific-and-other-technologies-by-the-police-and-wider-criminal-justice-system/.

[13] https://inews.co.uk/news/data-centre-power-home-glasgow-edinburgh-3973815?srsltid=AfmBOoqEecLve52HXQgU8rQLNQghgcf1A2wxwkyYzK1NxX8SEQ6IjCfU.

[14] https://www.fticonsulting.com/insights/articles/hidden-risk-data-centers-no-one-talking-about.

[15] New technology will allow citizens calling the national emergency number to upload data and livestream footage of events direct to police control centre—https://www.biometricupdate.com/202409/emerging-technology-and-emergency-surveillance-or-vice-versa.

[16] https://artificialintelligenceact.eu/the-act/.

[17] https://www.biometricupdate.com/202509/roc-highlights-facial-recognition-algorithms-consistent-performance-in-nist-tests.

When we talk about police biometrics like live facial recognition (LFR), we also talk about the importance of individual data rights and the need to advertise its use in advance. We talk about why the police must give everyone the right to opt out.

When they talk about new biometrics, the police talk about the additional capacity that LFR gives them to prevent sexual harm and monitor offenders on the register,[18] which is one of the AI use cases that get the highest level of public support[19]—but they also talk about the absurdity of letting everyone opt out in advance as it lets the wanted people opt out of being found by the very technology we bought for the police to find them.

When they talk about biometrics, the police talk about the additional, 'game-changing' capability that remote technology like LFR gives them, about how, with AI-enabled technology, they have a fighting chance against people who pay to watch children being raped on websites, and to identify, not just the perpetrators, but also the victims, some of whom have been ritually and anonymously tortured for lurid entertainment for years.

They talk about the new capability to enhance safety at resource-intensive public events like the Notting Hill Carnival.[20]

When they talk about biometrics, the police don't talk about CCTV because that's got nothing to do with biometrics, but they must if their councils are planning to load facial recognition software into existing street cameras in the name of crime prevention. When they talk about biometric technology, the police should be talking about the risks of hybridisation, about the differences between conventional and AI-enabled biometrics and about the risk of losing the ground they've recovered in public trust and confidence. And they need to start talking about their expectation that the citizen will keep feeding their insatiable surveillance appetite with private uploads, and the risk of the state becoming increasingly dependent on citizen-generated data captures and the aggregated biometric capability of the people.

When the police talk about biometrics, they talk about why they can search everything an arrested suspect owns, carries or controls, accessing bedrooms and medicine cabinets but not a biometrically locked mobile phone or device, even if it contains conclusive proof of the suspect's involvement in serious crime. And in

[18] https://www.biometricupdate.com/202506/is-this-the-perfect-use-case-for-police-facial-recognition.

[19] See CENTRIC survey on citizens' attitudes to police use of AI at https://aipas.co.uk/.

[20] https://www.biometricupdate.com/202508/from-notting-hill-to-denmark-hill-uk-police-facial-recognition-rejoins-the-carnival.

the UK they talk about the act of Parliament that tells the police what they must do when taking an impression of a suspect's boot print, yet says nothing about the mass capture and storage of body images, voice and iris patterns and other measures of biological identity.

When we talk about police biometrics, we have already begun to talk about its proliferation in the high street and how it's being seen as a smart way of alerting the police to burgeoning retail crime. We should instead be reflecting on how the use of biometric technology in shops and supermarkets is not 'police biometrics' at all, about how it's preventing some of the 20,000 reported assaults or 50,000 thefts in the UK each day and about the best way to measure the impact of biometrics[21] If we're going to talk about retail biometrics and policing, we should be talking about its preventative impact, about how age verification tools are helping shop-floor staff stop under-age purchases of alcohol and knives and other regulated items without feeling harassed or pressurised. We should begin to talk about biometric surveillance in shops as a health and safety issue rather than a security solution and how its greatest contribution to policing may be by avoiding the need to call the police at all.

We should probably talk about how some retailers— like their local councils—are already using plug-and-play biometrics as if they were just clever CCTV cameras, why using the technology to send images to the police is a crude under-utilisation of the technology to achieve a sub-optimal policing outcome,[22] and why a single enforced standard for commercial use of biometric surveillance is essential. While we're on the topic, we should talk about how partial regulation is no regulation, learning the lessons from other ID systems like vehicle number plates where being on the government's list of 40,000 registered suppliers in the UK is only effective at generating registration fees and does nothing to stop them selling plates illegally.[23]

When we talk about police biometrics in the round and look at how they can be regulated in the AI-enabled surveillance future, we might want to start with the only real question for any regulatory framework: does it give the public what they need?

---

[21] https://www.biometricupdate.com/202508/why-retail-needs-biometrics-the-cameras-arent-working.

[22] https://www.biometricupdate.com/202507/how-to-measure-success-in-biometric-security.

[23] Chapter 13.

Artificial Intelligence is bringing extraordinary biometric capability to policing, and it must bring extraordinary accountability with it.[24] The effective and enabling regulation of police biometrics needs an approach that is clear, consistent and comprehensive. When we talk about that, what we talk about must cover the *whole ecosystem in which any single element operates—this is because contamination or dilution of any part of a system contaminates or dilutes the whole.* Either demonising or fetishising a specific application like facial recognition is irrational, and expecting the law to move at the speed of creativity is fantasy.

When we talk about police biometrics, if we talk about possibilities and dependencies, about the increasing risks of *not* using AI-enabled solutions as well as those from deploying them, and about meeting people's legitimate expectations, we'll have a shot at preparing a planned response to a real-world future while there's still a chance. If we don't do this, then it won't matter what we talk about when we talk about police biometrics – it'll just be talk.

[24] Chapter 10.

The Policing Privacy Paradox 18

Changes brought to our lives by AI-enabled technology have created a range of challenges for the world of surveillance, as this book demonstrates. One such is a new—and so far, little noticed—phenomenon: the policing-privacy paradox.

Traditional police surveillance depends on perspective—where you are looking at something from will determine what you can see, and also what you will probably miss. When formulating surveillance strategy and policy, however, the police don't always remember this and can sometimes (ironically) overlook key aspects of the debate.

While the idea of physical perspective is becoming less relevant in the era of interconnected surveillance,[1] when trying to untangle the issues of AI-enabled surveillance, your vantage point can help. And one vantage point reveals a paradox for the police. The problem arises like this:

Wherever we live, we continue to adopt affordable surveillance technology (the possible) like home security apps and IoT devices[2] and build it into our everyday lives. Driven primarily by convenience,[3] millions of us now routinely access our sensitive personal data using biometrics such as facial images and fingerprints and allow access to commercial companies like banks. We are happy to allow airports to create templates of our faces and those of our nearest and dearest who are travelling with us. We use surveillance apps to monitor and protect our children and check visitors at the front door, in a way that has become commonplace. Some sectors like finance (Fintech), ticketing for sports and music

[1] Chapter 4.

[2] Chapter 2.

[3] Chapter 7.

F. Sampson, *How AI-Driven Surveillance is Changing our World*, Security Informatics and Law Enforcement, https://doi.org/10.1007/978-3-032-24005-7_18

events, and increasingly retail shopping,[4] have become so dependent on multi-factor authentication, Know Your Customer and digital identity technology that it's hard to imagine the pencil and paper bureaucracy of previous generations, a return to which now seems impossible. Government requirements like anti-money-laundering provisions (AML) and national digital identity mechanisms have also moved us onto these platforms.

Looking at what's now technologically possible, the innovation in our hands is changing our surveillance relationship, not just as consumers, but also as citizens. It is clear from their daily appeals for dashcam, smartphone or other image captures that police forces across the world now *expect* the tech-powered citizen at least to share the product of their private biometric surveillance capabilities (Sampson, 2025).

Panning across to the legal perspective (what must/must not be done), all this capturing and sharing needs to meet the relevant rules and regulatory framework. Given the greater consequences for the citizen when the police are involved, there may be different rules for the private use of biometric surveillance capabilities within the confines of our own lives when compared to corporate or even some state uses, but the citizen must nevertheless comply with the applicable laws governing data. Many countries treat the processing of biometric data such as facial images as requiring a higher level of scrutiny and regulation whoever is doing it, either at state or national/federal[5] Using biometric data is usually subject to more stringent regulation than more benign types of information like names and email addresses, while the use of AI systems to process it will up the ante even further.[6] In the future, data regulators and privacy bodies will need to maintain a close watch over this area if the boundaries between personal processing activities and police-directed surveillance activity become blurred.[7]

For now, it's enough to note how we are embracing innovative biometric technology and are even content to share some of our content with the state.

[4] G Kostka, L Steinacker and M Meckel, 'Between security and convenience: facial recognition technology in the eyes of citizens in China, Germany, the UK and the US (2021) 30(6) *Public Understanding of Science* 644.; Z Pingo and B Narayan, 'Privacy Literacy and the everyday use of social technologies' in S Kurbanoğlu,,*Information Literacy in Everyday Life* (ECIL 2018) 33.

[5] See, for example, lawsuit against Clearview AI alleging violation of Illinois residents' privacy rights under the Illinois Biometric Information Privacy Act (BIPA)—https://www.aclu.org/cases/aclu-v-clearview-ai.

[6] https://artificialintelligenceact.eu/high-level-summary/.

[7] Chapter 15.

Seeing us enthusiastically buying and using intrusive surveillance technology in our lives, checking bank accounts and getting into concerts, the police often assume that we would support *them* using new biometric surveillance capabilities like facial recognition technology (FRT). After all, many of the purposes for which we're deploying our AI-enabled capabilities (safety and security) are part of the overall strategic responsibilities being entrusted to law enforcement. It's legal and it works, so why not use it?[8] Then there's that stubbornly resistant law enforcement assumption that "if you've done nothing wrong, you've nothing to worry about".[9] From this perspective, opposing the police using the technology would seem *prima facie* paradoxical. But that's what has been happening in many countries.[10]

To understand what might be going on, we need to shift to the third vantage point. In terms of the *societal* perspective, we know people are very willing to share personal datasets and biometrics, surrendering sensitive personal data with private companies. In some circumstances they will also share it with the police.

But, as the philosopher Albert Camus observed, we live in contradiction; just because something is possible and permissible doesn't mean we will find every application of it acceptable.

Behavioural studies of online activity have shown that domain-specific privacy concerns do not sufficiently explain domain-specific privacy behaviour.[11] In other words, just because we are concerned about one privacy-sensitive issue does not always accord with our behaviour. This well-established concept, although still undeveloped, is known as the privacy paradox—as originally proposed, it has

---

[8] For the answer, see Chap. 11.

[9] The worst surveillance advice ever—see Chap. 9.

[10] See, for example, Australia—https://theconversation.com/australians-like-facial-recognition-for-id-but-dont-want-it-used-for-surveillance-new-survey-shows-235530; Canada—https://ccla.org/our-work/privacy/surveillance-technology/facial-recognition/, India—https://www.giga-hamburg.de/en/publications/giga-focus/digital-surveillance-and-the-threat-to-civil-liberties-in-india.

[11] B Brown, 'Studying the Internet Experience Publishing Systems and Solutions' (Laboratory HP Laboratories Bristol HPL-2001-49, March 26 2001); J Bryce and J Fraser, 'The role of disclosure of personal information in the evaluation of risk and trust in young peoples' online interactions' (2014) 30 *Computers in Human Behavior* 299; AD Ker, 'Decoding the Privacy Paradox: can we reconcile our privacy beliefs and behaviour?' (*The Privacy Issue* January 22, 2020). https://theprivacyissue.com/privacy-and-society/decoding-privacy-paradox.

found empirical support.[12] Observed online behaviour of customers and attitudes to biometric technology have also been recognised as a practical challenge in the financial technology sector, revealing a divergence between what customers believe about gaps or risks in security and their subsequent activity in spite of that belief (Rádulescu, 2018), or that we are always comfortable doing that in every (or even the same) context.

A good example of apparently paradoxical digital behaviour is the finding[13] that around one third of British citizens keep a photo or scan of their ID documents stored on their phones; that one in five store scans of IDs belonging to friends and family on their devices and nearly a quarter (23%) also keep bank account details in a non-encrypted format, knowing that this could lead to fraud and identity theft. More than half (51%) of respondents in the survey admitted to being uncertain about the security of their identity data stored on their devices or other digital locations, and some 15% had been victims of digital fraud the previous month.

Given such digital profligacy revealed by over 1000 surveyed responders (with the least careful being younger people), it would seem reasonable to assume we are unlikely to be *more* reticent about the police using our data improperly, but that's not the case. Other research by colleagues at the Centre of Excellence for Terrorism, Resilience, Intelligence and Organised Crime research (CENTRIC) shows how public attitudes to the police use of FRT and other remote biometric surveillance capability are ambivalent (Ezzeddine et al., 2023; Hill et al., 2022; Hintz & Dencik, 2016) and that internet users are more likely to disguise their real identity if they believe that the police are watching.

In the near future the policing-privacy paradox may deepen as there is a growing expectation—in the UK at least—that we will share our AI-enabled surveillance output with the state,[14] while the adoption of intrusive surveillance tools by the citizen does not appear to have inured us to the use of some new

[12] T Dienlin and S Trepte, 'Is the privacy paradox a relic of the past? An in-depth analysis of privacy attitudes and privacy behaviors' (2015) 45(3) *European Journal of Social Psychology* 285.

[13] According to a survey published by Thales https://www.biometricupdate.com/companies/thales-digital-identity-and-security.

[14] S Marsh 'UK owners of smart home devices being asked for swathes of personal data' *The Guardian* (7 September 2023). https://www.theguardian.com/world/2023/sep/07/uk-owners-of-smart-home-devices-being-asked-for-swathes-of-personal-data.

technology such as FRT by the state.[15] If anything, we have become increasingly *uncomfortable* with, and even resistant to its use. Added to that discomfort and resistance, the impact of early experimentation with new technology such as facial algorithms and continuing security issues such as serious police data breaches[16] makes any assumption of unequivocal citizen support much harder to corroborate. If the developing trend of sucking up citizen-generated images continues, the surveillance capability of policing will become increasingly *dependent* on those. Future state surveillance is already relying on the sharing of images and other data uploads from devices and systems operated privately by the technologically enabled citizen.[17] In any system, dependency equals risk and if the paradoxical digital behaviour of the citizen moves in a less contributory direction, that will have an (as yet unquantified) impact on operational policing.

The emerging evidence in relation to public attitudes towards the police use of AI generally is still being studied, and a 2025 survey at CENTRIC revealed some interesting and highly relevant findings.[18] The survey of 10,114 citizens across all four UK nations (ranging in age from 14 to 85+ years) found that the majority of respondents (61.3%) supported the use of AI in policing, provided it came with strong accountability, oversight, and technological and legal safeguards.

Some 64% of respondents believed that AI enhances police capabilities (compared to 12.4% who believe AI undermines them) and the survey showed a recognition of key advantages to AI adoption in policing, especially in the context of **national security, increased police effectiveness** in solving crime, and freeing up **more time and resources** to focus on high-priority offending.

At the same time, the survey identified significant citizen concerns about the future use of AI by law enforcement, with **lack of accountability,** potential over-surveillance or misuse of personal data and the police 'blaming' the technology for errors. Of particular significance is the finding that **online privacy** is a focus of concern for respondents, 39.6% of whom said they would be unwilling to contribute personal data for the training of AI systems for policing purposes,

[15] See e.g. Ada Lovelace Institute, *Beyond face value: public attitudes to facial recognition technology* (Report, September 2019). https://www.adalovelaceinstitute.org/case-study/beyond-face-value/.

[16] https://www.bbc.com/news/uk-northern-ireland-66578582; https://www.politico.eu/article/crime-victims-details-accidentally-included-in-police-foi-responses/; https://www.theguardian.com/uk-news/2023/aug/26/met-police-on-high-alert-after-it-system-holding-officers-details-hacked; https://news.sky.com/story/south-yorkshire-police-loses-nearly-three-years-worth-of-body-cam-footage-with-an-estimated-69-cases-affected-12945797.

[17] Chapter 5.

[18] https://aipas.co.uk/.

compared to only 9.3% who would be very willing to share their data. These responses are highly relevant to the context of the policing-privacy paradox, as are the regional variations that the survey found. The survey found differential attitudes to the use of AI across parts of the UK where policing is locally devolved (Scotland and Northern Ireland) yet the regulatory approach to public space surveillance remains grounded in data protection which is covered by a UK-wide framework.

Finally, the survey revealed that the public view of police use of AI is heavily contextual. Comparing 32 different scenarios, the survey showed how trust and acceptance of AI tended to be highest for those situations in which deployment carries low risks for the public and those where AI can reduce serious harms to vulnerable groups. In terms of police 'use cases', the highest level of acceptance was found in the use of AI to identify perpetrators of child sexual exploitation and terrorism.[19] Lowest levels of **acceptance were for the automation of core functions** like 999 emergency calls.

Public space surveillance is being shaped by and for a new digital relationship between the citizen and the police, a relationship that may eventually create a similar operational dependency to that emerging from fintech and personal security. The interaction on which the new citizen-police surveillance relationship is based will bring expectations on both sides, one of which is that we will continue sharing our privately captured images for the purposes of policing; another is that the police will be more accountable in their use of it. There is no reliable research about this emerging surveillance trend yet, but the evidence from the CENTRIC public attitudes survey highlights how **trust is conditional on the policing context**.[20]

The impact of the inevitable expansion of AI-driven surveillance capabilities available to both citizens and the police has yet to be worked through. From the CENTRIC survey, however, we know that, across the UK, three quarters (75.5%) said they want mandatory accountability processes in place *before* the police deploy AI, and fewer than 20% believe sufficient accountability for police use of AI exists.

Technological possibility and legal permissibility occupy much of the debate about AI-enabled surveillance, but failing to take the *societal* aspects (acceptability) into account risks frustrating the proper adoption of undeniably powerful technology available to policing. Paradoxical behaviour aside, the future of

[19] For examples of both, see Chap. 11.

[20] 54.9% trust the police to use AI in decision-support roles, whereas 36.0% trust AI to make policing decisions independently.

surveillance by the state will be elementally different from anything that has gone before—and that will need a wholly different regulatory framework.

References

Ezzeddine, Y., Bayerl, P., & Gibson, H. (2023). Safety, privacy, or both: evaluating citizens' perspectives around artificial intelligence use by police forces. *Policing and Society, 33*(7), 861.

Hill, D., O'Connor, C. D., & Slane, A. (2022). Police use of facial recognition technology: The potential for engaging the public through co-constructed policy-making. *International Journal of Police Science & Management, 24*(3), 325.

Hintz, A., & Dencik, L. (2016). The politics of surveillance policy: UK regulatory dynamics after Snowden. *Internet Policy Review, 5*(3), 1.

Rádulescu, A. (2018). Users' social trust of sharing data with companies: Online privacy protection behavior, customer perceived value, and continuous usage intention. *Contemporary Readings in Law and Social Justice, 10*(1), 137.

Sampson, F., (2026) 'The future of police surveillance and cameras', Chapter 16 in Future Policing, Technology and Transformation Huey, L., Ferguson, L. and Adams, I. (Eds), (2026) Routledge, New York

19 Pandemic Surveillance

In state surveillance, one person's protection is another person's oppression. Using iris scanning to confirm individual identity cuts the administrative costs of aid agencies providing refuge to some of the world's most vulnerable people.[1] At the same time, the availability of iris database records of people who helped British forces emerged as a source of urgent alarm following a change in government in Afghanistan.[2] Live facial recognition technology (FRT) can be used to alert staff if a vulnerable service user wanders beyond the perimeter of a care home, but it has also been used to alert the police if people of controlled ethnicity wander into zones from which the state has sought to exclude them.[3]

This oxymoronic interplay between emancipation and oppression is often summarised as a privacy vs security trade-off, but that isn't necessarily an accurate representation of the past—the future of surveillance is likely to involve balancing personal choice with the public interest.[4]

One of the great fears that many people harbour about surveillance is what one American legal researcher once termed *omniveillance* (Blackman, 2008) the state's ability to watch all of us, all of the time. At the core of that concern is the idea that one day we might all be covered by a single surveillance system

---

[1] https://impactmap.cam.ac.uk/iris-recognition-transforms-the-speed-of-refugee-aid-distribution/.

[2] https://www.hrw.org/news/2022/03/30/new-evidence-biometric-data-systems-imperil-afghans.

[3] https://www.aljazeera.com/news/2019/5/2/hrw-china-using-mobile-app-for-surveillance-of-uighurs.

[4] Chapter 7.

F. Sampson, *How AI-Driven Surveillance is Changing our World*, Security Informatics and Law Enforcement, https://doi.org/10.1007/978-3-032-24005-7_19

observing our every action, empowering the state to step in and enforce the law if we cross any lines that it has drawn. That day may be closer than we think. To see why, I will go back—as lawyers often do—to definitions.

> It is Thursday 21 May, the day before the end of Ramadan, and a new law has been passed by the state which orders the closure of all mosques. The legislation also covers all other religious denominations and effectively says 'no person who wishes or, as a matter of their religion is required, to attend a collective act of worship at their mosque, church, synagogue, temple or chapel is permitted to do so'.

On the face of it, this may seem like some 'dystopian' fiction beloved by surveillance sceptics, or an act of a new authoritarian regime, but it is neither: it was a real law and it was passed in the UK, which is surprising until we look at the year—it was 2020.

Since that year the word 'pandemic' has been used to mark what was a particular point in time, when the latest variant of the COVID virus was detected. But the adjective pandemic comes from the Greek *pan demos* meaning 'across all people'. And the ubiquity of technology in our lives suggests that we may be in a surveillance pandemic.

The COVID-19 outbreak provides more than just a handy linguistic device for a discussion of surveillance. The conditions accompanying the outbreak were extraordinary,[5] and the response from governments was equally so.

In the case above a deeply concerned citizen[6] challenged the UK government's emergency legislation[7] which stopped any gatherings for worship unless they were for funerals, broadcasts or urgent support. In dismissing the case, the judge outlined what were 'truly exceptional circumstances, the like of which has not been experienced in the United Kingdom for more than half a century' when we were faced with a virus that 'is a genuine and present danger to the health and well-being of the general population'.[8]

That the High Court was even considering a government order banning public worship in the UK in the twenty-first century is telling enough.[9] The risk of

---

[5] And probably broke all previous records for use of the word 'unprecedented'.

[6] *R (on the Application of Hussain) v Secretary of State for Health and Social Care* [202] EWHC 1392 (Admin).

[7] The Health Protection (Coronavirus Restrictions) (England) Regulations SI202/350.

[8] *Hussain loc cit* at 19.

[9] Even writing that sentence causes me to revisit the case and check, for the umpteenth time, that this was really happening.

serious infection was truly global—it was *pan demos*—as were the measures needed to combat its spread, measures that included increased state surveillance.

The exigencies of the COVID-19 pandemic required emergency, temporary measures, some of which involved segregation, restriction on movement, assembly and attendance at school, as well as the ban on public worship challenged by Mr Hussain.[10] Preventing people from doing things that for the past half a century had been cherished legal entitlements and fundamental freedoms was never going to be straightforward. As became uncomfortably clear by some events that followed,[11] even at a time of global health emergency, people cannot necessarily be relied on to follow all the rules all of the time. Once orders that severely limited the movement and association of the population had been properly enacted, they had to be enforced, and that means a step-change in surveillance. For a while, local police forces became agencies of *health* enforcement.[12] Even viewed against the scale of the emergency, the extraordinary powers for policing were seen as being unacceptably broad in their scope and discriminatory in their execution, with the civil rights group Liberty claiming that the surge in stop and search, chaotic communication and overzealous enforcement led to misuse, the inappropriate issuing of fines and other behaviour that severely undermined the British model of policing by consent.[13] Examples of police surveillance and enforcement tactics included the use of drones to identify picnickers in the UK,[14] 15.5 million citizen stops by April 2022 and the disproportionate use of 'audits and inspections' across Europe (Aitkenhead et al., 2022). In a move away from the Peelian principles, the police in the UK began issuing on-the-spot fines known as fixed penalty notices (FPNs). Whatever the tactics deployed and their consequent impact on reducing public confidence in the police (Mynenko & Ditcham, 2022), all police responses nevertheless required a high level of vigilance and observance over the entire population—it was a period of literal *pandemic* surveillance in both senses of the term, watching everyone to constrain a global disease.

[10] The impact of which is still being scrutinized by the UK parliament -https://covid19.public-inquiry.uk/.

[11] https://www.newstatesman.com/science-tech/coronavirus/2021/12/timeline-all-the-lockdown-breaches-by-politicians-and-government-employees-so-far.

[12] Strictly speaking, enforcers of health laws—https://www.college.police.uk/research/projects/policing-pandemic-role-enforcement-securing-compliance-coronavirus-regulations.

[13] https://www.libertyhumanrights.org.uk/issue/pandemic-of-police-powers-liberty-reveals-scale-of-misuse-of-police-powers-under-lockdown/.

[14] https://www.bbc.co.uk/news/uk-england-derbyshire-52055201.

Tackling the spread of COVID-19 across the entire population needed the enlistment of the population, and this often went beyond people acting as volunteers assisting with essential local services to contributing to the surveillance capability of the police. The rapid expansion in citizens helping the police by uploading images and their geolocation data has been tracked through several other chapters and, in the future, AI-enabled surveillance systems (including the marshalling of citizen surveillance power) will be increasingly deployable to tackle the threat of infection, civil or military emergency. Where temporary emergency measures are put in place, it is critical to ensure that they are exactly that—impermanent and used only to the extent necessary to counter the threat. Sometimes that is easier said than done and not every emergency surveillance tool can be put back in the box.

In addition to the more predictable features such as a blurring of law enforcement and health enforcement, there were some other less obvious biometric and surveillance lessons from the COVID-19 pandemic.

One biometric legacy for policing came from the need to minimise the use of arrest powers. To avoid unnecessarily bringing people into close contact with each other in custody, the police in England and Wales used a scheme of voluntary attendance (VA). Over the period of the emergency provisions in the UK, the use of VA increased dramatically (Sampson, 2023). While this was a good thing for constitutional lawyers—arrest is highly intrusive on citizens' freedoms—it was a bad thing for biometric capability. In a country where the power to take biometrics of a suspect is directly linked to their arrest, it's potentially calamitous. As a result of VA many suspects in criminal investigations passed through the hands of the UK police over a period of around 18 months without having their fingerprints or DNA taken.[15]

The imposition of 'contactlessness' prompted revisions to everyday practices in non-police settings and some organisations became very creative in their responses.

As Biometrics and Surveillance Camera Commissioner I had not envisaged spending much time talking about school meals. However, once governing bodies started arguing for the introduction of FRT to take payment for meals on the basis that it was a COVID-preferable alternative to conventional methods, the discussion widened.

[15] How many and to what effect is not yet known.

While not a typical surveillance scenario, the school dining hall saga is a good illustration of how organisations can avoid making a dog's' breakfast of AI-enabled technology. The basic arguments to consider before installing biometric surveillance go like this:

1. Biometric data is highly sensitive personal information and attracts specific protection from the law.[16] Additional sensitivities that arise when dealing with biometrics of minors are well known and need special treatment.[17]
2. Before processing biometric data, you must consider whether it is justified. By definition, temporary emergency conditions—such as a global pandemic—call for solutions that should not apply for long. Permanent long-term solutions are the opposite.
3. You must also consider the proportionality of the solution and whether the same outcome can be achieved by other less intrusive solutions—in the contactless payment scenario it should be clear what other already-contactless payment arrangements you have rejected and why.
4. AI-enabled technology is in an almost permanent *beta* state, meaning new applications for the same technology will appear in the future. Once an organisation has persuaded itself of the legitimacy and proportionality of using biometric surveillance technology for one function (paying for school meals to ensure that the right meal is served to the right student), the same rationale will probably apply to other similar payment scenarios (school trips, ticketed events, issue of sports and musical equipment etc.).
5. Once installed, the new technology no longer needs the old infrastructure. Some of your savings to pay for it will come from dismantling the tills, the capability for handling money, cashiers and so on. There's no way back.
6. With all less intrusive options, the next proportionality test for people after you will be a shoe-in. There will be a *stronger* business case for subsequent applications of the intrusive technology than in the original emergency case, virtually guaranteeing function creep.[18]

The end state in the school scenario is a dining hall camera that cannot only recognise who's paying for what, but also who's playing lead in the school play and a generation of adults for whom omniveillance is the default.

[16] https://gdpr-info.eu/art-4-gdpr/.

[17] Something that has directly shaped the regulatory framework of police surveillance in the UK—*S & Marper v The United Kingdom* [2008] ECHR 1581.

[18] A widely held concern about AI-enabled surveillance technology—see Chapter 10.

Lawyers generally come at this as a balancing act, and balancing individual rights isn't always commutative—A × B isn't necessarily the same as B × A. Nor is it a see-saw activity where you put all your emphasis on one end to counteract the other. With surveillance technology, the governance and regulatory challenge is more dynamic, and the balancing is gyroscopic. It requires constant adjustment, weighing what can be done technically with what must or mustn't be done and what people will tolerate or expect to be done in their name and neighbourhood. Even in emergencies that will vary considerably.

Pandemic threats warrant pandemic solutions but does that mean every global emergency threatening the lives of millions justifies near-total state surveillance? That might be arguable for a time-limited extreme situation where the threat to the global population is so high it merits whatever surveillance we can muster. But what about other crises? Many scientists have testified to the immediate and irreversible threat to humanity presented by global warming and climate change. If that isn't that *pan demos* what is? Does that apply to very local, practical issues like installing emission zone cameras to penalise 'dirty' engines? When the purpose is to avoid the death of our planet, where are the limits on state surveillance capabilities? What would a *disproportionate* use of ANPR cameras to enforce Ultra Low Emission Zones look like? The COVID-19 pandemic shows we may need to think about 'proportionality' as more than a purely legal concept.

We also need to consider societal expectations. Public support for police use of AI attenuates as you move along the continuum from terrorism and serious crime to automating calls for help.[19] Where does using drones to monitor picnic areas feature on that public acceptability scale? Policing activity routinely involves interference with constitutional rights such as the right to peaceful enjoyment of property, privacy of home life and correspondence, and gathering in worship. But at times of imminent threat, defeasance can be beguiling. Switching off the constitutional firewall might be easy but at what enduring cost?

Technological developments in biometrics and surveillance have meant that our capability and capacity to prepare for, respond to and recover from global crises have increased beyond anything our forebears might have realistically imagined; at the same time, this has created dependencies and vulnerabilities on a similar scale.

The risks from temporary measures becoming permanent incursions by the state during times of emergency are well documented (see, e.g., Lowande & Rogowski, 2021). As Levitsky and Ziblatt observed, 'Citizens are more likely

[19] Chapter 10.

to tolerate and even support authoritarian measures during security crises—especially when they fear for their own safety' (Levitsky & Ziblatt, 2019). One of the many surveillance challenges in future emergencies will be thinking about how the balance between responsible intrusion, accountable regulation and societal expectation can be resumed.

More prosaically, pandemic emergencies can offer a 'lazy defence' for policies of internal convenience, particularly where the temporary measure saves money. This can bring unintended consequences, as was the case with the UK, where arrangements to avoid filling police cells and common custody areas produced a significant increase in VA that *didn't return to pre-COVID levels* after the emergency period.[20] Because VA doesn't give the police the power to take DNA samples, the number of enrolments on databases plummeted. At a time when the criminal justice system is in a state of 'slow motion collapse',[21] it is not difficult to envisage an emergency 'early release scheme' being an attractive response to a national health emergency.

The coalescence of critical features—the need to monitor the entire population and a growing capability to achieve it—makes 'omniveillance' a subject of universal relevance. We have probably not witnessed the last global-level infection—when the next one hits, the realities of AI-enabled surveillance will be *pan demos*.

References

Aitkenhead, E., Clements, J., Lumley, J., Muir, R., Redgrave, H., & Skidmore, M. (2022). *Policing the pandemic*. Police Foundation London.

Blackman, J. (2008). Omniveillance Google, privacy in public, and the right to your digital identity: A tort for recording and disseminating an individual's image over the Internet. *Santa Clara Law Review, 49*, 313–392.

Levitsky, S., & Ziblatt, D. (2019, January 12). Why autocrats love emergencies. *New York Times*.

[20] Sampson ibid.

[21] Matt Dathan 'Criminal justice system in state of slow motion collapse'. The Times 22 October 2025 https://www.thetimes.com/uk/crime/article/criminal-justice-system-in-state-of-slow-motion-collapse-xl3986qz3.

Lowande, K., & Rogowski, J. (2021, June 15). Executive power in crisis. *American Political Science Review, 114*(4). Cambridge University Press.

Mynenko, K., & Ditcham, K. (2022). *Public confidence in the police: A new low for the service.* RUSI.

Sampson, F. (2023). Annual report of the biometrics and surveillance camera Commissioner 2022–23, 30 October 2023 House of Commons Library.

20 Shrinkworld and the Hunt for the Red Tesla

The police often talk metaphorically about 'shrinking' a suspect's world. Policies or tactics that will 'leave suspects no place to hide' are frequently cited by chief officers to reassure their communities and put their adversaries on notice. The availability of AI-enabled surveillance capability at times when they will be expected to use it[1] now empowers the police to shrink worlds, not just for suspects, but for all of us.

The state's ability to watch all of us, everywhere, all the time has been a consistent feature of novels and Hollywood thrillers for decades, but AI-enabled technology, coupled with the way we have started to use it ourselves, is making omniveillance (Blackman, 2008) a reality[2]—and in ways that we might not have noticed, still less predicted.

The focus of surveillance literature—fact and fiction—tends to be a proliferation of visible infrastructure supported by intrusive powers for the state to support it. Banks of police cameras and blinking red lights everywhere. While some of this picture is accurate, the previous chapters have shown that the evolutionary path of AI-enabled technology is heading somewhere quite different. More state surveillance doesn't necessarily mean more cameras (Sampson, 2025)—as I have said many times, surveillance is no longer about where the police put their cameras; it's about what they do with the images from everyone's cameras. That feature alone has super-charged the state's ability to watch us and changed our surveillance relationship with the police forever.[3]

[1] Chapter 11.

[2] Chapter 4.

[3] Chapters 7 and 15.

F. Sampson, *How AI-Driven Surveillance is Changing our World*, Security Informatics and Law Enforcement, https://doi.org/10.1007/978-3-032-24005-7_20

This is not all the state's doing. We are shrinking our own worlds, and we are much more conspicuous than we've ever been before. Our use of technology in our homes, work and private lives has made us very visible to, and our habits have made us findable by, governments, people and corporations far beyond our own shores. By signing up to data sharing protocols in the pursuit of convenience, we are surrendering elements of what we once called 'privacy' to global companies for whom personal information is a form of digitally mineable rare earth. Self-promotion via social media and our eagerness to show everything is denuding us and eroding our digital coastlines at an extraordinary speed. In surveillance terms, we've made the state's job of watching us a lot easier—so easy, in fact, that any residual notion of privacy relies on the state *not looking*.[4]

There's no reliable measure for how heavily 'surveilled' places or people are,[5] and therefore it's hard to assess the dimensions of our digital worlds, but we can see where AI-enabled surveillance is heading. Accelerator events are speeding up the shrinkage—operational successes using biometric technology to catch sexual predators or avert terrorism and dismantle organised crime will help build public trust and confidence, in both the technology itself and in the ways the police use it. But the most likely trigger for blanket police surveillance will probably not be a law enforcement scenario at all. To understand why, we need to look around us.

The surveillance community finds prediction irresistible, with some pinning great hopes on the power of AI-enabled technology to watch what hasn't even happened[6] Much of this actuarial policing fad doesn't stand up to scrutiny yet and we need to more than just casting the digital twigs to see how they fall. Helpfully, there are practical ways of looking down the strategic highway.

Anticipatory planning is essential in setting strategy and budgets for public bodies like the police,[7] and academics use Delphi techniques to scope situations and foresee context (see e.g. Jorm, 2025).

A more flexible (and fun) approach is the one pioneered by Peter Schwartz (1991), whose book, *The Art of the Longview*, introduced the use of scenario planning to track business trends and directions. Schwartz argues that most big changes in our environment don't just appear unannounced; they evolve (albeit very quickly at times) out of what's already here.

---

[4] Chapters 15 and 16.

[5] Chapter 4.

[6] Does looking at the yet-to-come count as 'surveillance'?

[7] https://www.cipfa.org/services/networks/finance-advisory-network/events-material/improving-financial-strategy-forecasting-and-decision-making.

Surveillance professionals will immediately grasp this concept. In the policing world, the issuing of a BOLO (be on the lookout) 'reveals' things by making them stand out. Take the example of a red Tesla. Once you're asked to keep an eye out for one, red Teslas are everywhere. They were already there, of course, but they were indistinct among all the other cars until you were actively looking for them. In psychological terms, the instruction to look for them made the Teslas 'figure' rather than 'ground' (Peterson & Skow-Grant, 2003).[8]

With hindsight, anyone can see that the strategic indicators of the next big thing were already here: the billion-dollar question is how do we know what we're supposed to be looking out for right now?[9] The trick is being able to identify what the revelatory features of the future are before picking them out in the crowd. And in the setting of state surveillance, we have some clues.

This book tracks some of the emerging features, mapping where AI-enabled surveillance is taking us and vice versa; its chapters may contain red Teslas. While making no claims to prophetic accuracy, by reflecting on what's now possible and how we are using technology as a society, we can try to catch a glimpse of what AI-enabled state surveillance might look like and prepare for its arrival. That has been at the heart of this book.

Adopting Schwartz's methodology, we have highlighted some key certainties and uncertainties already visible today. In the AI-enabled surveillance world known certainties include the growth in citizen-generated surveillance and the changing surveillance relationship between each other and the state, rapid expansion in remote biometrics and ID technologies, edge computing, Deepfakes etc. By positioning them against known uncertainties—legal and regulatory frameworks, paradoxical online behaviours, applicability of existing norms to AI-enabled outcomes, greater tolerance of intrusion in proportion to perceived threats—we can pre-empt the different surveillance environments and conditions based on immediate evidence rather than fictional foreboding.

Now we can look at scenarios in which all this comes together and model possible police surveillance futures. Scenario planning, as pioneered by Schwartz, involves identifying macro-level changes—economic, biological or technological—in your environment and positing scenarios around it. However unlikely or far-fetched they might seem now, the scenarios can throw up trends, patterns of behaviour and directions of technological experimentation. Plotting the more significant ones, then assessing their certainty, can expose what's coming and reduce the element of surprise. Scenarios can uncover the red Teslas.

8 The same thing happens if you're told to look for someone bald, or a post box or smoke.

9 I.e. How do we know the red Teslas hold the answer and not, say, the green bicycles?

Taking the macro-level technological changes in their ability to shrink worlds, we can see peacetime policing scenarios that demand the greatest level of population visibility (in terms of percentage of people who need to be monitored and the spatial and temporal coverage necessary). In the past 50 years, the one requiring and receiving surveillance saturation came not from crime and disorder but from disease: the global COVID-19 pandemic.

International responses to the outbreak demanded a degree of global omniveillance which is consonant with the meaning of the word 'pandemic' itself.[10] The sheer breadth of the 'pandemic' threat called for pandemic surveillance to police it. The sociobiological responses to the pandemic brought new national and local policing issues with it, and left some problems behind. Relationships between communities and their police were strained by the novelty and enormity of the event, and there was a blurring of law enforcement and health enforcement functions. There were other interesting evolutions (see Jahankani et al., 2021) and lessons from that period show that, on the one hand, people blatantly breached the emergency laws and, on the other, the policing response went beyond what was necessary and proportionate to enforce them, with examples of overreach and undue zeal.

Since 2019, technological developments in biometrics and surveillance have increased capability and capacity to prepare for, respond to and recover from global crises. One of the certainties in our surveillance future is that the next global health crisis will be monitored (at least in part) by AI-enabled technology. What might the realities of that look like for the police and their communities?

For the purposes of illustration, the following scenario takes the form of an imagined report covering Operation Sentinel, the UK police response to a new pandemic:

> Police in the UK have introduced a suite of 'health and safety focused tools', and tactical responses to tackle the new pandemic. Using AI-enabled technology to manage a spike in calls for police service while minimising the risk of infection from contact with officers and staff, police leaders have drawn on experience from the COVID-19 era. Reviews of policing highlighted the benefits of reducing arrests and detention, while at the same time noting that DNA and fingerprint records of suspects were not checked for over 18 months –a lag that lasted several years causing the enrolment onto biometric databases to plummet.
>
> AI help pods have been installed in streets and shops to replace police station help desks and users will be able to speak to a police chat bot to get up to date information about the disease, lockdown rules and restrictions. The chat bots can also

[10] From the Greek *pan demos* meaning across the population.

receive reports (anonymously if preferred) together with images and data feeds from the public and have multi-lingual translation and British and American sign language capability. To access the pods users must first confirm their identity by submitting their biometrics (face, fingerprint or iris) to register and sign in. The police say this is necessary to improve accuracy in their records, reducing fake reports while increasing biometric databases (to be used 'strictly for policing purposes'). The bots will be using DeepSee, a search algorithm that will allow them to compare sign-in data against all police records including historical crime scene markers, photographs and other fields for the purpose of 'finding vulnerable and missing people and potential witnesses'. DeepSee also searches other government biometric databases in the UK and those of international partner agencies.

Local police leaders have asked people to be patient and vigilant. 'We have limited resources which are very stretched at this time' said Commander Saul, Sentinel's chief officer. 'Obviously we can't be everywhere, so we are relying on the community to be our eyes and ears'. With most people having access to smart devices and ways of capturing evidence, the police are relying on the aggregated capability of the community as a vital part of their response to the disease. Saul explained, 'During periods of enforced confinement and quarantine, we are asking everyone to use their devices—doorbells, dashcams, wearable cameras, home security etc.—to look out for their neighbour. If you are concerned that someone is in need of assistance or is breaching the health rules, you should collect any reliable data from a safe distance (photographs, sound recordings) and upload it to the police infection control centre. They don't have to be high quality or professional but please make them as clear as you can, taking care to make sure that a person's full face can be seen or that number plates are readable where a vehicle is involved. If necessary, our operator can take temporary control of your device remotely using special software'.

Local authorities have also been keen to emphasise how their investment in AI-enabled technology will dramatically improve their ability to keep people safe. Some authorities have set up public space monitors by retrofitting biometric capabilities in their existing street CCTV systems, others have taken measures that are less obvious. For example, people making essential visits to some public buildings will be scanned by biometric sensors to collect health data (temperature, heart and pulse rate) which can be as unique as a fingerprint in identifying individuals and may be uploaded to the cloud for analysis and processing by the relevant monitoring service. During the emergency period, this data may be accessed by law enforcement.

Councils and other local services have also set 'geofences' around public areas which sound an alert if individuals come closer together than the relevant distance as set by the Chief Medical Officer. Leaders of the local authorities have clearly taken some pride in these new capabilities. Councillor Li Grahn said 'We have invested in state-of-the-art technology. This time there won't be secret prayers or staff parties—we're taking people's safety very, very seriously'. In place of the pavement stickers and notices on furniture used during the COVID-19 pandemic, local authorities are using geofencing technology that will, as Grahn puts it, 'stop any arguments about loopholes in the law' and make sure people 'follow the spirit as well as the letter' of the lockdown arrangements. Some areas will be closed to everyone and, as breaching a geofence without lawful authority or reasonable excuse is a criminal offence, the police will have access to all monitoring data.

There are concerns about the level of intrusion from these technological arrangements. Geofencing relies on access to personal devices via GPS, Wi-Fi, cellular networks, or Bluetooth. The emergency regulations make it mandatory for people to enable location services on their phones and give relevant permissions to the geofencing monitoring service. Failing to do so is a criminal offence. Some employers already use geofencing to track employees' movements at work, monitoring assets like vehicles and electronic devices—they have signed data sharing protocols with the police and local authorities.

Care homes were a critical feature in the previous pandemic, and they will have mandatory geofencing, both around their external perimeters and internally to segregate residents inside the buildings.

If there is a need for police attendance at any residential premises, Commander Saul says they will be greatly assisted by AI technology. For example, the police response may involve sending an *i*-Ball®, the latest AI-enabled surveillance vehicle, to investigate conditions outside and detect the presence of any geofence or other alert capability. Saul advises that 'This is an autonomous AI-operated police device that is focused firmly on the safety of everyone involved'. Personal protection drones (nicknamed Heli-coppers) may also be despatched to observe the scene from a safe distance using 360-degree compound eye lenses. The drones, which are 'invisible' to the naked eye, can remotely access internet-connected devices or home security cameras to assess the likelihood of there being infected or vulnerable people inside a building, or any substances or devices that present a risk. They may also have records of residents' lockdown movements including Cell Site Location Information from mobile phones, GNSS and vehicle telemetry from the onboard car computers. Asked about their intrusive impact, Saul said, 'They're a new way of gathering information safely and accurately. Please be reassured that, provided you are complying with the rules, you have nothing to be concerned about even if you can't see them'.

Using what they say is an 'aggregated picture', before sending officers the police will assign a VIX (Volatility Index) rating to an address. The rating (between 1.0 and 7.0) will be available to officers and the policy says: 'online behaviour markers or other reasons for significant concern (court judgments, non-crime incidents etc.) of any occupants can affect your VIX rating'.

The *i*-Ball® will have the VIX rating and has AI-enabled surveillance capabilities including always-on video and audio communication with the control centre. Advanced features include a lip-reading function and a translation capability. Capturing facial images of anyone at the scene, the *i*-Ball® will carry out near-instant facial, tattoo and biometric matching and will be equipped with a remote enrolment device to take fingerprints. People will be able to apply to have these records deleted once the emergency situation has ended. It will also be able to issue fixed penalty notices (FPN) to anyone it suspects of breaching the health regulations.

In light of the threat to national security, officers have the authority to examine any devices on premises. All computers, tablets, laptops (including those used by children) will need to allow access to their content and officers will interrogate any home management device (Alexa, Siri etc.) to identify any prompts that users have given it. They may also access any internet-connected or remotely operated domestic devices like fridge freezers, thermostats and heating boilers, as well as satellite tracking devices fitted to things like sports watches, trainers and household pets. If it is necessary to remove any of these (including the pets?) for further examination, owners will be issued with a digital receipt and can collect their property by arrangement using the Help Pod.

The police surveillance policy published on their website states that 'any images that are shared with us (whether by you, friends, family, employers, customers or others known or unknown to you) may be subject to retrospective facial matching against police and government records and OSINT (open-source intelligence) checks across social media and other local sources of data'. If you have a UK passport and/or driver's licence, officers will have the photographs and, this being a health emergency, they may have access to personal medical records and vaccination history.

Concerns have already been raised about the longer-lasting impact of this technology and the risk of 'mission creep'. The Information Commission has received several complaints about homeowners using their Wi-Fi routers to scan neighbour's movement through the walls and counting the number of people on the premises, while privacy groups have pointed out that sourcing citizen-generated surveillance allows the police far more intrusive capability than if they tried to operate the technology themselves, and bypasses the regulatory frameworks designed to ensure accountability. The *i*-Ball® is made by a Chinese technology company in Ghuanyzhou and this been raised by several MPs as a national security issue with the Surveillance Camera Commissioner. Some cyber experts are also alarmed at the risks of hacking the mega-databases of biometric information that the government has now compiled and have pointed to the police's poor track record of data breaches.

A statement from the Home Office said 'These are extraordinary times and call for extraordinary measures. Our overarching focus is to keep people safe and return life in the UK to normality as soon as possible. All policies have undergone careful scrutiny and are compliant with the prevailing temporary health regulations. Further information can be downloaded from the website at the **P**olice **and Em**ergency **I**nformation **C**entre (PANDEMIC)'.

The above situation is, of course, fictitious but all the technology is real.[11] Each of the capabilities referred to is already here and the scenario reveals an ability for saturation surveillance. The locus of AI-enabled surveillance has shifted from places to people and the scenario envisages near-total access to almost every

[11] For a list, see Appendix.

category of personal data about the entire population at any time. If that's not omniveillance, what is?

How it will be used during the next pandemic is unknowable, but the world-shrinking technology is already available to the police and will pose some hard questions. If they can save thousands of lives and help contain a global pandemic with some of this technology, how could the police ethically *not* use it?[12] How realistic is it to expect the surveillance capability and the records it created to be 'switched off' once the emergency has passed? Data could be deleted, but many records will be needed for future public inquiry—and deleting it attracts allegations of a lack of candour which goes directly against the government's policy of transparency and public accountability. Once public funds have been invested in technology, the case for retaining some of its functions must be considered and those cases might include other emergencies and major investigations. Who will decide what they are?

We have been enthralled by stories about the government watching us for several generations—most famously by George Orwell whose name has become synonymous with oppressive state surveillance. But looking ahead isn't always fictitious, and fiction isn't always right. Looking back to see where we've been is also important in building the available 'intelligence picture'. That picture is based on what we know and what we don't, with the most worrying part being the complete blanks, the famous Rumsfeldian Unknown Unknowns.[13] With AI there's not a lot to look back at as we haven't been here before,[14] so we can only look ahead and most of what we can see is made up of Unk-Unks.

In looking for the red Teslas, we know three things:

1. Technologically, the police can see into the lives of most of the population and expansion of measurables like gait, heart rate, brain patterns is replacing biometrics with 'zoemetrics'.[15]
2. Legally, the conditions legitimising the use of that technology are proportional to the risk—pandemic tools can be used to meet pandemic risk.

---

[12] The ethics of surveillance is a vast subject—for a starter see Kevin Macnish (2025) The Ethics of Surveillance: an introduction, Routledge.

[13] The things we don't even know we don't know—as opposed to Known Knowns and Known Unknowns, made famous by US Secretary of Defense Donald Rumsfeld during the run up to the Iraq war in 2002http://news.bbc.co.uk/1/hi/magazine/7121136.stm.

[14] Chapter 1.

[15] The measurement of all manifestations of life.

3. Societally, support for authoritarian police surveillance is highest during emergencies.[16]

In the AI-enabled future, one of the many challenges will be guarding against the permanent and irreversible incursion of surveillance is deemed necessary during a temporary crisis. Someone will need to make sure that what is technologically possible and legally permissible is *only* used in a way that remains societally acceptable, maintaining the critical equilibrium between innovation, regulation and expectation.

References

Blackman, J., (2008). Omniveillance, Google, privacy in public, and the right to your digital identity: a tort for recording and disseminating an individual's image over the Internet. *Santa Clara Law Review* 49, 313–392

Jahankani, H., Kendzierskyj, S., & Akhgar, B. (Eds.). (2021). Data privacy and security: Some legal and ethical challenges in information security technologies for controlling pandemics (pp. 109–134).

Jorm, A. (2025). *Using the Delphi method to achieve expert consensus.* Palgrave Macmillan.

Peterson, M. A., & Skow-Grant, E. (2003). Memory and learning in figure-ground perception. *Psychology of Learning and Motivation, 42*, 1–35.

Sampson, F (2025). Policing, AI and the New Surveillance Relationship. In: Groves, M and NG, Yee-Fui, (eds.) Automation in Governance Theory, Practice and Problems. Bloomsbury Publishing.

Schwartz, P. (1991). *The Art of the Long View.* Doubleday/Currency.

[16] Citizens are more likely to tolerate and support authoritarian measures during security crises, especially when they fear for their own safety (Levitsky and Ziblatt).

Appendix

Technology from Chapter 20

Geofencing—https://litum.com/geofencing-technology-for-security/?keyword=&creative=&gad_source=1&gad_campaignid=20575637987&gbraid=0AAAAADODlI7nsjwG43Y8aBaMCZU5xclbs&gclid=EAIaIQobChMI2YnFua_fkAMVuZhQBh2veg9iEAAYAyAAEgKkrfD_BwE

***i*-Ball®**—https://www.biometricupdate.com/202412/china-testing-ai-powered-police-ball-that-looks-like-a-battle-droid

Lock Down Patrol Robot—https://www.theguardian.com/world/2021/oct/06/dystopian-world-singapore-patrol-robots-stoke-fears-of-surveillance-state

Hornet Drone—https://www.insta360.com/product/insta360-sphere

Officer-Initiated Facial Matching App—https://www.south-wales.police.uk/police-forces/south-wales-police/areas/about-us/about-us/facial-recognition-technology/faqs-for-operator-initiated-facial-recognition-app/

Open Source Intelligence (OSINT)—https://www.police-foundation.org.uk/2021/10/open-source-intelligence-the-cinderella-of-the-investigative-family/

Biometric Sensors—https://www.electromaker.io/blog/article/biometric-sensors-2022-buyers-guide?srsltid=AfmBOop030saeYi7DCENbZgZ5hw5vDkUWLytgTq0k2_G_RPHA31RcZw6

WiFi Routers Seeing Through Walls—https://www.popularmechanics.com/technology/security/a42575068/scientists-use-wifi-to-see-through-walls/

F. Sampson, *How AI-Driven Surveillance is Changing our World*, Security Informatics and Law Enforcement, https://doi.org/10.1007/978-3-032-24005-7

Pet Trackers—https://tractive.com/?google30&gad_source=1&gad_campaignid=23201970622&gbraid=0AAAADfJtI_RRf-BD2aDvMkML4wDktyMb&gclid=EAIaIQobChMIrKvouq7fkAMVIZFQBh3g7BwCEAAYAiAAEgIwRPD_BwE

Running Shoe Trackers—https://www.runnersworld.com/gear/a20853776/track-your-run-with-smart-footwear/

Golf Ball Trackers—https://www.rfidlabel.com/rfid-chip-golf-balls-cut-search-time-by-80-and-focus-on-your-swing/?srsltid=AfmBOooVA40ZTp_1i4W8SryT-O0UhVw298Ruq3VCIu8sxzSt2G8z5Tga

Retrofitting Council CCTV With Facial Recognition—https://www.lbhf.gov.uk/news/2025/09/new-cctv-technology-help-met-police-fight-crime

Face recognition glasses—https://glassalmanac.com/meta-reveals-facial-recognition-plans-for-smart-glasses-in-2026-why-it-matters-now/

Remote measurement of temperature, heart and pulse rate—https://www.sciencedirect.com/science/article/pii/S2468227623000959

Citizen uploads to police control rooms—https://blue-link.co.uk/about.html

Robotic police dogs—https://nypost.com/2024/11/08/us-news/robot-dog-seen-patrolling-donald-trumps-mar-a-lago-estate/

Iris scanning to register for public services—https://www.cam.ac.uk/stories/iris-recognition-technology

Index

F. Sampson, *How AI-Driven Surveillance is Changing our World*, Security Informatics and Law Enforcement, https://doi.org/10.1007/978-3-032-24005-7

GPSR Compliance

The European Union's (EU) General Product Safety Regulation (GPSR) is a set of rules that requires consumer products to be safe and our obligations to ensure this.

If you have any concerns about our products, you can contact us on ProductSafety@springernature.com

In case Publisher is established outside the EU, the EU authorized representative is:

Springer Nature Customer Service Center GmbH
Europaplatz 3
69115 Heidelberg, Germany

Batch number: 10460812

Printed by Printforce, the Netherlands